T0093853

Synthesis Lectures on Mechanical Engineering

This series publishes short books in mechanical engineering (ME), the engineering branch that combines engineering, physics and mathematics principles with materials science to design, analyze, manufacture, and maintain mechanical systems. It involves the production and usage of heat and mechanical power for the design, production and operation of machines and tools. This series publishes within all areas of ME and follows the ASME technical division categories.

Arif Sirinterlikci · Yalcin Ertekin

A Comprehensive Approach
to Digital Manufacturing

 Springer

Arif Sirinterlikci
Industrial and Manufacturing Engineering,
School of Engineering, Mathematics,
and Science
Robert Morris University
Moon Township, PA, USA

Yalcin Ertekin
Department of Engineering Leadership
and Society
Drexel University
Philadelphia, PA, USA

ISSN 2573-3168   ISSN 2573-3176   (electronic)
Synthesis Lectures on Mechanical Engineering
ISBN 978-3-031-25353-9   ISBN 978-3-031-25354-6   (eBook)
https://doi.org/10.1007/978-3-031-25354-6

This Springer imprint is published by the registered company Springer Nature Switzerland AG
The registered company address is: Gewerbestrasse 11, 6330 Cham, Switzerland

Preface

This book draws a comprehensive approach to digital manufacturing through computer-aided design (CAD) and reverse engineering content complemented by basic CNC machining and computer-aided manufacturing (CAM), 3D printing, and additive manufacturing (AM) knowledge. The reader is exposed to a variety of subjects including the history, development, future of digital manufacturing, a comprehensive look at 3D printing and AM, a comparative study between 3D printing and AM and CNC machining, and computer-aided engineering (CAE) along with 3D scanning. Applications of 3D printing and AM are presented as well as multiple special topics including design for 3D printing and AM (DfAM), costing, sustainability, environmental, safety, and health (EHS) issues. Contemporary subjects such as bio-printing, intellectual property (IP) and engineering ethics, virtual prototyping including augmented, virtual, and mixed reality (AR/VR/MR), and industrial Internet of Things (IIoT) are also covered.

The book is structured in a way that Chap. 1 introduces digital manufacturing, its role in product design and development, and the historical evolution of rapid prototyping into AM while comparing two of digital manufacturing's main actors, 3D printing and AM, and CNC machining. Following Chaps. 1, 2 prepares the reader for the CAD subjects relevant to 3D printing and AM as well as CNC machining. As Chap. 3 is designated to CAM and CNC machining, Chaps. 4–6 present the pre-processing, processing, and post-processing stages of 3D printing, respectively, including the technologies involved and their associated materials. Applications of 3D printing such as rapid prototyping, rapid tooling, and AM (formerly referred to as rapid manufacturing) are presented in Chap. 7. Chapter 7 also includes non-industrial use cases of 3D printing. IIoT and CAE are the subjects of Chap. 8, also including digital (virtual) twins. Reverse engineering methodology and associated technologies including 3D scanners, IP laws and ethics, and non-industrial use cases wrap up the content of Chap. 9. Chapter 10 encompasses cost modeling and estimation as well as business proposal development, sustainability and EHS issues for 3D printing and AM, and multiple realities content including AR/VR/MR for IIoT and industrial training. The book concludes with Chap. 11 of possible project ideas in product development and manufacturing, reverse engineering, and automating 3D printing and AM.

Each chapter comes with in-practice exercises and end-of-chapter questions, which can be used as home-works as well as hands-on or software-based laboratory activities. End-of- chapter questions are mainly of three types: review questions which can be answered by reviewing each chapter, research questions which need to be answered by conducting literature reviews and additional research, and discussion questions. In addition, some of the chapters include relevant problems or challenges which may require additional hands-on efforts. Most of the hands-on and practical content is driven by the authors' previous experiences. Authors also encourage readers to help improve this book and its exercises by contacting them.

Moon Township, PA, USA   Arif Sirinterlikci
Philadelphia, PA, USA   Yalcin Ertekin

Acknowledgements

Arif Sirinterlikci thanks his parents Suzan and Celal, his wife Aleea, and his daughter Selin for their love, encouragement, and support, along with his uncle Mel for inspiration. His collaboration with the following Robert Morris University (RMU) professors led to the development of parts of this book: Dr. P. Badger, Professor and Department Head of Science, Dr. D. Short, Associate Professor of Environmental Science, Dr. E. Erdem, Associate Professor and Coordinator of Industrial and Manufacturing Engineering, Dr. W. Joo, Associate Professor of Biomedical Engineering and Department Head of Engineering, Dr. R. Carlsen, Associate Professor of Mechanical and Biomedical Engineering, Director of RMU Center of Innovation and Outreach, Dr. J. Al-Jaroodi, Professor and Coordinator of Software Engineering, and Dr. N. Kesserwan, Assistant Professor of Software Engineering. Dr. L. Monterrubio, Associate Professor and Coordinator of RMU Mechanical Engineering also deserves credit for his background in CAE and the help he provided for this book. Dr. A. Stefaniak and his team at the Centers for Disease Control (CDC)/National Institute of Occupational Safety and Health (NIOSH) have been instrumental in EHS and recycling work done in 3D printing and AM, and documented in this book. PA Department of Community and Economic Development (PA DCED) and America Makes provided financial support for some of Dr. Sirinterlikci's initiatives which are mentioned in this book in addition to CDC/NIOSH. Companies like Microsonic, Schroeder Industries, Union Orthotics and Prosthetics, and GE Additive have also been excellent collaborators for presenting invaluable research/development and consequent learning opportunities. The following RMU students are some of the recent contributors to the work presented in this book and deserve credit: Ms. L. Farroux, Ms. A. Wolfe, Ms. M. Orlando-Jepsen, Mr. C. Bill, Mr. J. Pham, Mr. B. Krider, Mr. Z. Wilson, Mr. M. Kutzmonich, Mr. J. Thomas, Mr. S. Rakus, Mr. Z. Giovaniello, and Mr. P. Connolly. Other earlier contributors can be seen in the references. And finally, Dr. Sirinterlikci is grateful to Mr. G. Cottrell, RMU Laboratory Engineer, and Mr. G. Yarmeak, former RMU Laboratory Engineer for their contributions.

Yalcin Ertekin thanks his parents Insirah and Rafet, they are the greatest inspirations in his endeavors. He also thanks his wife Lale, for her love, encouragement, and patience, as well as his children, Aylin and Ozan, for their love, understanding, and continued support. He would like to extend his appreciation to numerous colleagues at Drexel University including the former and current heads of the Department of Engineering Technology, Dr. V. Genis and Dr. J. Tangorra, respectively. They truly strived for best teaching and learning environment to develop several of the learning materials that went into this book. He also wants to thank his colleagues Dr. R. Chiou and Dr. I. Husanu, who provided many collaboration opportunities in developing engineering technology mechatronics research and teaching laboratories equipped with modern CNC machines, industrial robots, and AM equipment. Drexel University's Steinbright Career Development Center is acknowledged for providing financial assistance to students in co-operative education at Engineering Technology mechatronics labs. One of those co-op students was Mr. M. Olean, whose contribution is greatly acknowledged. The following undergraduate students also were inspirational in developing the learning tools and materials: Mr. S. J. Konstantinos, C. Ruiz, E. Jenkins, C. Eger, G. Sappington, M. Kowalski, D. Garg, J. Haas, B. Cohen, C. Tsai, Y. Akyondem. The author also wants to acknowledge the National Science Foundation (NSF) for providing several research and teaching grants supporting the author's research and development.

Contents

About the Authors

Arif Sirinterlikci Ph.D., CMfgE is a University Professor of Industrial and Manufacturing Engineering at Robert Morris University (RMU) School of Engineering, Mathematics, and Science (SEMS). At RMU, he served in different administrative roles including Head of Engineering Department, Associate Dean of Research and Outreach at RMU SEMS, and most recently as the Senior Director of RMU Center of Innovation and Outreach (CIO). He holds BS and MS degrees, both in Mechanical Engineering from Istanbul Technical University in Turkey and his Ph.D. is in Industrial and Systems Engineering from the Ohio State University. He has also been a Certified Manufacturing Engineer (CMfgE), awarded by the Society of Manufacturing Engineers (SME), since 2016. Dr. Sirinterlikci was actively involved in SME serving in its Journals Committee, and Manufacturing Education and Research (MER) Community Steering Committee. He also served as an officer of the American Society for Engineering Education (ASEE) Manufacturing Division between 2003–2011 including its Chair. He is one of the original grant writers of the winning proposal/founders of the U.S. Government's America Makes Institute. His teaching and research interests lie in 3D scanning, 3D printing and additive manufacturing, product design and development including biomedical device development, industrial automation and robotics including mixed reality and Internet of Things (IoT) applications, computer-aided engineering (CAE) in manufacturing processes, and entertainment technology. Dr. Sirinterlikci has over 130 conference papers and publications, trade and scientific journal articles, and book chapters to his record, a few being recognized as best paper nominees by the ASEE Manufacturing and Engineering Technology Divisions and his Rapid Prototyping Journal environmental, health, safety issues (EHS) paper winning a highly commended paper award from the Emerald Literati Network in 2016. He is a teaching professor who tries to follow the teacher/scholar model by engaging his students in his research and bringing his research to his classroom. Three of his students recently won the Italian Technology Awards (ITA) in two consecutive years, 2018 and 2019, and 4 groups of his students earned provisional patents.

Yalcin Ertekin Ph.D., CMfgE, CQE is a Clinical Professor in the College of Engineering, Department of Engineering Leadership and Society at Drexel University, Philadelphia, and serves as the Associate Department Head for Undergraduate Studies for the Engineering Technology program. He received his BS degree from Istanbul Technical University in Turkey, an MSc in Production Management from the University of Istanbul, and an MS in Engineering Management, and MS and Ph.D. in Mechanical Engineering from the University of Missouri-Rolla. Dr. Ertekin has also been a Certified Manufacturing Engineer (CMfgE), awarded by Society of Manufacturing Engineers (SME) since 2001 and a Certified Quality Engineer (CQE) awarded by American Society for Quality (ASQ) since 2004. In addition to positions in the automotive industry, Dr. Ertekin has held faculty positions at Western Kentucky University and Trine University. In 2010, he joined Drexel University's College of Engineering as an associate clinical professor. He has been instrumental in course development and the assessment and improvement of the Engineering Technology (ET) curriculum, including integrated laboratories, project-based learning, and practicum-based assessment. Dr. Ertekin serves as the faculty advisor for the student chapter of the Society of Manufacturing Engineers (S058) and is a member of the College's Undergraduate Curriculum Committee. Involved in research, Ertekin has received funding from the National Science Foundation (NSF), private foundations, and industry. His research has focused on the improvement of manufacturing laboratories and curricula and the adoption of process simulation into CNC machining, additive manufacturing, and other traditional manufacturing practices. His areas of expertise are in CAD/CAM, manufacturing processes, machine and process design with CAE methods, additive and subtractive manufacturing, quality control, and lean manufacturing.

Introduction

1

Manufacturing is one of the critical factors determining the livelihood of a nation's economy and wealth of its people. In parallel, innovation is equally imperative. Both intersect often as the product-based innovation relies on new developments in manufacturing processes, their tooling, and materials utilized while production innovation improves the performances of manufacturing enterprises greatly, making them competitive in the rapidly changing global markets.

As shown in Fig. 1.1, manufacturing industries have come a long way from mechanization and steam power driven predecessors of the eighteenth century (the First Industrial Revolution, also known as Industry 1.0), introduction of electrical power, assembly lines, and mass production in the nineteenth century (Industry 2.0), followed by semiconductor-based electronics, digital computers, and automation late twentieth century (Industry 3.0). Industry 4.0 is allowing manufacturing machines, their components including their tooling and sensors to be a part of the cyber world, based on the cyber physical systems and Internet of things (IoT) concepts and communicate with other machines and computers [1, 2].

If its impact is neglected, the Industry 4.0 may initially seem like a small part of the fundamental computer-integrated manufacturing (CIM) concept of integration of all of the engineering, manufacturing, and business functions of an enterprise [3]. As depicted in Fig. 1.2, the CIM concept encompasses *computer-aided design (CAD) activities* such as geometric modeling, engineering analysis (usually carried out as computer-aided engineering or CAE), followed by a design review and evaluation as well as automated drafting tools. CIM also includes a variety *of computer-aided manufacturing (CAM) tools for manufacturing planning* such as cost estimation, computer-aided process planning (CAPP), numerically controlled (NC) part programming, computerized work studies, materials requirements planning (MRP), and capacity planning, along with other *CAM*

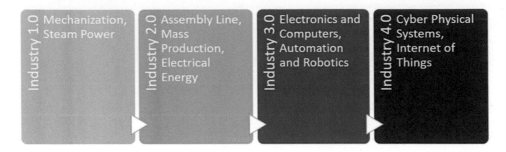

Fig. 1.1 Evolution of manufacturing industries

tools for manufacturing control: process monitoring, process control, shop floor control, and computer-aided inspection (CAI). *CIM business components* are geared for computerized functions like order entry, accounting, payroll, and customer billing. In the Industry 4.0 environment, a machine builder can monitor the health of one of its computer-controlled machines from a whole world away, while the manufacturing enterprise employing the equipment can track the performance of not only that machine, but also the overall system. If the collected data and information is collided with an artificial intelligent (AI) application (big data analytics), we reach a truly *smart manufacturing system* that adjusts itself by learning. This system will have high levels of adaptability in its supply chain through connectivity including in its production operations, offers multi-scale dynamic modeling and simulation, intelligent automation, sensor networks and information security [2, 4]. If the production operations are environmentally friendly or sustainable, and the power is transmitted to/from a smart grid, the smart manufacturing system will also be *green or sustainable*.

This book draws a comprehensive approach to digital manufacturing through CAD and reverse engineering content complemented by basic computer numerically-controlled (CNC) machining, 3D printing and additive manufacturing (AM) knowledge.

1.1 What is Digital Manufacturing?

Digital manufacturing is simply an integrated approach utilizing the digital design of a part, product or tooling for making it, via numerically controlled machines such as 3D printers or CNC machines. If the end goal is to make a prototype quickly, the process is called *rapid prototyping*, no matter what the nature of the process is; additive, subtractive, formative, or a combination of the two or three—hybrid. If the end goal is to obtain an end- product or tooling, then the process is referred to as *rapid manufacturing* and *rapid tooling* respectively. Even though both concepts finish with a different outcome including a prototype, end-product or tooling, the premise is that they are identical in nature. *Additive manufacturing* (AM) is a newer term, which replaced the term rapid

Fig. 1.2 Cornerstones of a CIM system [3]

manufacturing, becoming widely used since most of the processes are additive, however it does not reflect any subtractive, and formative processes itself.

Digital design of a part, product or tool is today obtained by computer graphics via CAD tools. On the contrary, companies may also be forced to 3D scan existing old pre-CAD era components to obtain a digital image of them. Figure 1.3 explains in detail the workflow for digital manufacturing, through two different (CAM) pathways: *3D printing/AM or CNC machining*. In the case of 3D printing/AM, once a CAD file is generated, it can be saved in intermediate data formats (such as STL or OBJ) before it can be sliced for layered processing for tool path generation complemented by process parameter information including (i.e.) deposition speeds, process temperatures. Support structures are also added to the numerical code along with other printing data during pre-processing, canceling the work-holding requirement with clamps, fixtures, or jigs. In the case of CNC machining, once the digital model of the part is completed, each machine tool operation is defined by the tool involved, geometry to be cut, and again the process parameters including machining process type, feeds, speeds, depths of cut, number of passes, and step-over distances alike. Once the toolpath for each operation is completed, it is play-backed (back-plotted) and verified in a virtual 3D environment, and post-processed for NC code generation. In the meantime, work-holding set-ups (instances) have to be prepared for the machining operations.

Digital models of future products or tooling are also utilized as *virtual prototypes* in CAE environments for design review and evaluation. Figure 1.4 shows a Moldflow application for processes, where a future part is being analyzed for process cycle-time [5]. These types of analyses' (*CAE for manufacturing processes*) results can be used, during the development stage, in optimizing part and tooling designs in addition to optimizing process parameters, all for defect free good parts made in short cycle-times with minimal costs. Once the part, process, and tooling designs are optimized, the resulting digital models can be used in the digital manufacturing process such as CNC machining or 3D printing of tooling. A variety of similar CAE tools are available including Magma or Procast for casting simulation, PAM-STAMP for sheet metal processing, and DEFORM for forging. In a contrasting example, PAM-CRASH software has been helping automotive engineers in understanding crash performance of their designs since 1985, and is based on similar Finite Element Analysis (FEA) tools utilized in CAE environments [6]. In a similar fashion, the *traditional structural CAE analysis* can be used in estimating the strength of a forged automotive component. In today's digital manufacturing, digital models of systems are also employed in discrete-event simulation of digital factories, tracking and analyzing material, part/product, worker movements/safety, storage and machine utilization (*CAE for manufacturing systems*) along with CAM tools like Fanuc's HandlingPro for OFF-LINE robot programming as shown in Fig. 1.5.

Furthermore, *digital twin and thread* concepts are starting to play a greater role in understanding dynamics of physical assets such as machines and help improve their performances with continuous improvements of their designs and work settings. Digital twin is model of a physical twin (asset) utilizing various means to improve the modeling and simulation of that asset. The digital twin given in Fig. 1.6 acquires real-time data from the IoT sensors associated with the asset (i.e. a pump) to help adjust the simulation model of the asset. In addition, expert human input, data from similar machines or past usage of the same machine can be utilized in improving the impact of the digital twin. With the help of the digital twin concept, the companies can understand and optimize the product design and its performance, detect and isolate faults and take corrective action, and generate maintenance/service schedules for prevention [7, 8]. As depicted in Fig. 1.7, a digital thread describes an asset (i.e. a product)'s data/information flows throughout its product lifecycle. The digital thread can also address networking protocols, security, and

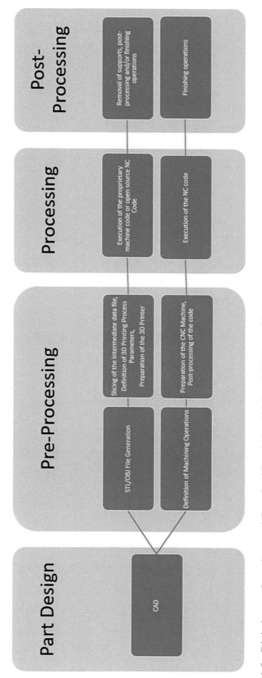

Fig. 1.3 Digital manufacturing workflow for 3D printing/AM and CNC machining

Fig. 1.4 Moldflow cycle-time
analysis for design evaluation

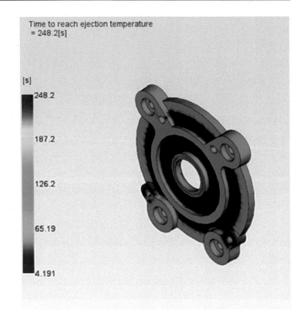

Fig. 1.5 Fanuc's HandlingPro
robot OFF-LINE programming
software

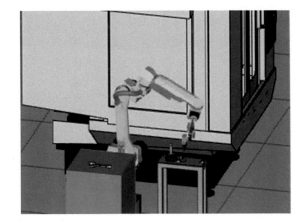

standards relevant to those data/information flows. Typically, the digital thread connects
digital twins in groups to present the right information/data at the right place within the
CIM structure. For example, the product design activity in Fig. 1.7 will produce a product definition, its relevant information and data to be connected to the production process
planning activity to be followed by the execution of the production processes and as-built
records [9].

The CAD/CAE/CAM integration has been carried out mainly by the human factor
manually over the last few decades as the major software companies tried to integrate

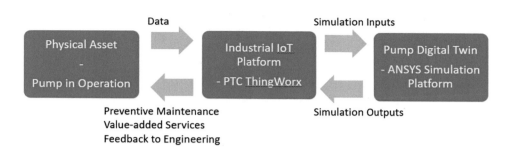

Fig. 1.6 A digital twin application—creating a digital twin for a pump [8]

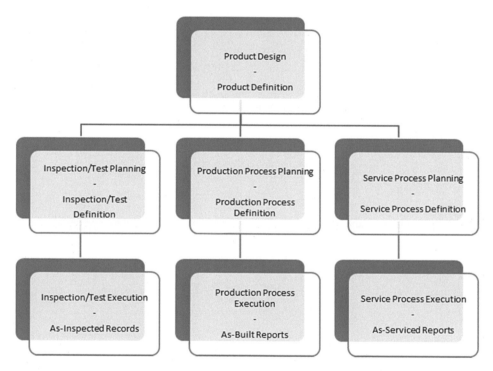

Fig. 1.7 Digital thread of a product [9]

CAE/CAM modules into their CAD software. On the other hand, newly developed software tools like nTop Platform incorporate CAD/CAE/CAM integration through a single platform for designing, analysis, and preparation for digital manufacturing, also enabling team collaboration through iterations using engineering notebooks. Alike other similar tools, nTop Platform is aiming to reduce the reliance to the STL geometry, and is based on a 3D Implicit Geometry Kernel, with spatially varying (lattice) structure generation ability and topology optimization [10].

1.2   Role of Digital Manufacturing in Product Design and Development

According to Ulrich and Eppinger [11], successful product development is reflected by resulting product quality (including functionality, reliability. maintainability, easy handling, and aesthetic appearance), product cost, and development lead-time for response to the market, impacted by the developmental cost and capability. Product development is an interdisciplinary effort requiring almost all functions of the enterprise. However, *marketing, design and manufacturing* are found at the center of the development activities. Marketing takes a critical role with identification of customer needs and market segments, design functions define the form and function of the product meeting the needs of customers, and manufacturing making products of good quality and lowest possible costs. Just like any engineering design and development process, the product design and development *starts with planning and concept development, followed by system-level and detail design, and completed with testing and refinement, leading to production ramp-up (start-up)* [11].

- Planning: Takes place before the actual product development process is approved by the company. This stage includes adherence to corporate strategic plans and assessment of technology developments and market objectives. The target market for the new product, business goals of the development project along with key assumptions and constraints are determined.
- Concept Development: The market needs are identified, along with alternative product concepts to be evaluated for satisfying the customers' needs. The concept is defined by its form and structure, and its functions are driven by the product specifications. A competitive benchmarking and economic justification of the project are also prepared at this stage.
- System-level Design: The proposed product is broken down into subsystems including assemblies, subassemblies, and individual components, in the form of a tree diagram or an assembly diagram.
- Detail Design: Encompasses a complete specification of the geometry, material selection, dimensions/tolerances of each individual part, including the make/buy decisions. A process plan is established along with the CAD files describing the whole product including its assemblies, subassemblies, and working drawings of parts to be made along with production tooling.
- Testing and Refinement: Involves the fabrication and evaluation of multiple prototypes of the selected product design. Early *alpha prototypes* are form and fit checked, and tested for appropriate functionality internally. Later *beta prototypes* are also evaluated internally, but also tested by prospective customers in focus groups for improving the performance of the product while identifying any quality issues. Table 1.1 explains

Table 1.1 Prototype and early product's adherence to actual product and production requirements [11]

Product and process features versus prototype stage/product	Intended geometry	Intended materials	Intended production processes	Intended assembly processes
Alpha prototype	x	x		
Beta prototype	x	x	x	
Product design at the production ramp-up	x	x	x	x

the adherence of three stages of the prototype and product's adherence to the intended final product geometry, materials, production and assembly processes.

- Production Ramp-Up: Happens before switching to actual production. The product is made by utilizing the intended production and assembly system. This is for working the final kinks out of the system and training the workforce. Any remaining design problems are also resolved. Additional field-testing and evaluation may also be conducted in parallel by utilizing the selected customers.

The role of digital manufacturing in product design and development process is clear. Rapid physical and virtual prototypes are involved in checking design (fit/form/function), materials, tooling, and process issues. They, especially the virtually prototypes, can be employed early and often to help improve the prospective product and its production environment, even though the physical prototypes mainly belong to the testing and refinement stage. Early identification of design flaws saves the developer time and money, allowing shorter developmental lead times, leading to products of better quality and lesser costs. Rapid tooling can be also utilized in making prototype tooling during the development or production tooling for the intended production system. Finally, rapid manufacturing can be a part of the intended production system.

1.3 Historical Evolution from Rapid Prototyping to Additive Manufacturing

The layered nature of the 3D printing process is similar to the deposition and etching of the integrated circuit (IC) fabrication methods, originating in the late 1950s. However, Hideo Kodama from the Nagoya Municipal Research Institute is considered as the pioneer of 3D printing [12]. In 1981, he envisioned using a laser to cure a photopolymer in his layered rapid prototyping system concept, applied for a patent, but never completed the process due to lack of funding. Alain le Mehaute, Olivier de Witte, both associated with

Alcatel, and Jean-Claude Andre, of the French National Center for Scientific Research (CNRS) also worked on the idea of laser curing liquid monomers into complex shapes, but ended-up abandoning their project after filing a patent in 1984 due to lacking funding as well [12]. In the same year, Charles Hull presented the idea of curing layers of photopolymers by using an ultraviolet (UV) lamp to make small, custom parts to his employers. Hull applied for a patent in 1984 and earned it in 1986 [12]. He named the process *Stereolithography (SLA)* and employed a UV laser rather than a lamp [13]. His new company 3D Systems released its first commercial printer SLA-1 in 1988 [12]. In the same year, a patent was filed by Carl Deckard, a University of Texas undergraduate student, on the *selective laser sintering (SLS) process* [12]. His first machine, Betsy, was aimed to prove the concept of utilizing lasers for sintering powdered plastics in making 3D shapes, with no attention to print details or quality. A startup, DTM, was formed for commercial development of SLS machines [13]. As the SLS patent was being processed, a patent was filed for another process called *fused deposition modeling (FDM)* by Scott Crump, the co-founder of Stratasys [12]. The FDM process is based on extruding heated thermoplastics and depositing them on a build plate in layers. The FDM patent was received in 1992. As the rapid prototyping boom continued in the 1990s and early 2000s, new processes were being developed during many more attempts for commercial success. Unfortunately, majority of the new processes could not achieve commercial success and the larger well-established companies absorbed some other relatively successful ones. Some of the noteworthy transactions include 3D Systems acquiring DTM in 2001 and Z Corp in 2012, and Objet and Stratasys merger in 2012. Z Corp was known for its *binder jetting process*, originally used with starches and later with gypsum making full color parts, and Objet owned the *polyjet process*, which was based on jetting out photosensitive material to be flashed by a UV light source in solidifying a whole layer.

In 2005, another major milestone in 3D printing history occurred. Adrian Bowyer launched the *replicating rapid prototypers* (RepRap) project for development of open-source 3D printers, leading the way to machines like Darwin or Mendel [14]. However, most importantly, it opened the door for low-cost machine development that resulted in Makerbots and the affordable machines of today, with a price tag of $200 or lower. The RepRap concept was originally envisioned for 3D printers replicating themselves by printing identical printers in their entirety. However, up to date, the concept mainly worked on printers making structural parts such as brackets, print head enclosures, gears. Even though some commercial printers now have the capability to print electronics, most of the RepRap machines are not able to do so. Since Stratasys still holds the trademark for the process' name fused deposition modeling, a new term for the FDM process was needed—*fused filament fabrication* (FFF) and is being applied to non-Stratasys commercial machines and RepRap kits.

Another milestone that made an impact not only in 3D printing but also in technological innovation happened in 2009 with the launch of the *Kickstarter project* [15]. Since then, many 3D printing projects were crowdsourced. As the low-cost alternatives were being developed, they lacked the quality and functionality the high-cost industrial machines offered. Thus, until recently *service bureaus* were compensating for the accessibility issues faced by small and medium size companies. With the *3DHub concept* individuals who had 3D printers also served as suppliers to the other individuals and companies for a short span that recently ended with low-cost and low-end printers becoming affordable to almost everyone as they were also being improved. Expiration of patents also led to development of competition as companies like 3D systems made small SLA printers like Projet 1200 available, only to be matched by Formlabs' Form printers.

Over the years, 3D printing's reach has extended simply from being a rapid prototyping technology into being a true rapid tooling, functional prototyping, and rapid (additive) manufacturing one. As the new processes developed in the last two decades, the material space has also exploded. Early 3D printers processed paper, starch, limited number of thermoplastics, waxes, and epoxy resins. Thus, they fell short in ceramic, metal, and composite printing areas. However, new processes currently are not only capable of printing with ceramics, metals and composites, but also do it in a much less costly and improved way.

Rapid tool making has been one of the main applications of 3D printing, through *indirect and direct tooling methods.* Direct rapid tooling refers to 3D printers directly making the molds, cores, and dies. In the case of the Voxeljet machines or ExOne's S-systems, based-on the binder jetting process, these printers make sand molds and their cores for the sand-casting process [16, 17]. One of the authors utilized the SLA process and DSM Somos Nanotool material, an epoxy resin filled with nano ceramic particles, for printing injection molding inserts and molded polycarbonate (PC) parts in the inserts [18]. An early indirect rapid tooling effort occurred at the Ohio State University in 1994 where an FDM 1600 machine was utilized in printing wax models to be converted to metal injection molding inserts through the investment casting process. At the time the idea was novel, but shell mold cracking issues caused the project team to CNC machine the metal inserts, greatly elongating the lead-times. This approach is now successfully carried out today repeatedly, reducing the lead-time and cost of tool making or metal parts compared with more expensive alternatives including traditional methods. In another indirect tooling application, the same author employed stainless steel prints obtained with EXONE's R2 machine as master patterns to make silicone rubber molds to cast RTV copies of the metal print using polyurethane [19]. The project was aimed for replicating historical artifacts with lesser costs, in addition to preserving the artifacts in digital (from a 3D scan). Printing jigs, fixtures, and cutting tools have become common over time adding to applications of rapid tool making.

Applications of 3D printing has been transitioning from rapid prototyping for product development to *rapid (additive) manufacturing* of industrial parts and end-use products as well. In a major attempt to migrate its manufacturing operations into 3D printing, GE Aviation acquired Morris Technologies, a leading supplier of contract additive manufacturing services, and its sister company, Rapid Quality Manufacturing in 2012. Both companies were under a contract to produce components for the LEAP jet engine being developed by CFM International, a 50/50 joint company of GE and Snecma (SAFRAN) of France [20]. In the spring of 2016, GE opened its Center for Additive Technology Advancement (CATA) as its applied R&D and solutions center for its businesses including Aviation, with an initial investment of $40 million and 50 workers [21]. These moves were followed by acquisition of the Swedish Arcam, the *electron beam melting* (EBM) company in late 2017, and purchasing 75% stake of German Company Concept Laser, the owner of the patented LaserCUSING technology for powder-bed-based laser melting of metals (*direct metal laser melting/DMLM aka selective laser melting/SLM*) mid-December 2016 [22]. Concept Laser's 3D metal printers process powder materials of stainless steel and hot-work steels, aluminum and titanium alloys and for jewelry making with precious metals [23]. Over the last couple decades the metal, ceramic, and composite markets have been dominated by European companies, ExOne being the only American company in metals printing. Currently GE Additive offers materials and extensive development consulting for various industries, besides owning a good part of the metal space with the purchase of the original EBM and DMLM companies. According to GE Additive, "it is dedicated to the further development and transformation of the industrial sector with software-defined plants as well as networked, adaptable and forward-looking solutions." [23]

Over the years the applications and their products have expanded from manufacturing industries including automotive and aerospace to other industrial products in lighting, clothing, jewelry, customized medical, dental and human products, a wide variety of food areas including pasta, pizza, cheese, pancake, chocolate as well as the construction industry with 3D printed buildings and bridges. Following section highlights some of these additions and new trends:

- A good portion of 3D printing work currently lies in hobbyist and educational space, and students are being introduced to this technology at very early in their educational careers, thanks to the RepRap movement.
- Medical applications have become more important with the development of software tools such as Mimics and 3-Matic for converting DICOM (Digital Imaging and Communications in Medicine) images from a medical scan into CAD files for printing. Thus, design and manufacturing of custom implants became a reality. Early work in late 1980s to early 1990s at UT Austin included using tricalcium phosphate (TCP) powders in bone printing. Similar materials like hydroxyapatite (HA) were used as scaffolds for tissue regeneration. Since then, patients are receiving full lower jaw implants, full hip implants, and the US Food and Drug Administration (FDA) gave 510(k) clearance for the world's first long-term plastic implant Osteo-Fab Patient Specific Cranial Device (OPCSD) manufactured by Oxford Performance Materials employing biocompatible PEKK (poly-ether-ketone-ketone) polymer and the SLS process. Over the years, start-ups focusing on *bioprinting* (printing living tissue

with encapsulated stem cells—bioink) started making an impact as companies like Envisiontec continued to market their bioplotters successfully. Researchers have been working on printing organs, meat or leather made from animal stem cells.

- Composite printers with reasonable costs like Markforged's OnxyPro are now allowing introduction of glass, carbon, and Kevlar additions to be introduced to their filament via the *continuous filament fabrication* (CFF) process for making reinforced parts. One of the authors has been developing polymer-metal composites for the FFF process, and now similar composite filaments are freely available for any FFF machines.
- Metal printing has also seen major changes in the last few years. Companies like Desktop Metal, Markforged, Rapidia introducing new machines and processes with lower costs than EBM, DMLM/SLM. Desktop Metal's three step process includes extrusion of metal with a polymer binder and a ceramic layer separating the part and the supports, debinding, and sintering. On the contrary, Rapidia has a machine extruding a metal/water paste, and sintering the print to a strengthened condition.
- As the 3D printing materials development accelerates in multiple directions, there is also a strong need for *blended and recycled plastic development* to help allow recycling of many plastics, which are occupying our landfills and waters.
- Often scientific and engineering developments are used to harm humanity, and 3D printing found its place in controversy. There have been concerns of individuals printing illegal guns and other objects, which can be used as weapons.
- Current trends also include development of *true 3D printing* where 3D geometry is directly built without a layered structure. A layered printing method in reality is made through a 2 and ½ D controlled process, where x- and y-axes are controlled simultaneously for each layer and the print head is moved in the z-axis direction in between two layers. This causes anisotropic material properties in the prints.
- Another new development is *4D printing* where a 3D printed structure is modified over time (making the time the 4th dimension) and the process 4D printing. The 3D printed structure may be responding to its environment when exposed to a catalyzer forcing a change in it. Future applications will include a good number of biological or chemical processes, which may lead to changes in the 3D prints over time.
- Stereolithography (STL) geometry has been dominating the 3D printing world. However, an STL file is not as accurate as its CAD model since it is made from triangular facets, losing geometric information during the conversion. There have been other geometries like PLY, 3MF and OBJ being utilized to *address the shortcomings of the STL geometry*, i.e. OBJ for addition of color and free form curve and surface representation.
- *Material database development, FEA simulation and design of experiments (DOE) methods* for optimization of materials, parts, and process parameters along with *closed-loop process control* have been some of the focus areas of the researchers.
- *Printer clusters or farms* are able to make more parts via 3D printing, and with help of *print management software* and inclusion of set-ups like Ultimaker S5 Pro Bundle

with multiple spool capacities lights out (automated) 3D printing is becoming a reality. These footsteps may also allow manufacturers to achieve *mass customization* in the near future.

• *Integration of 3D printing* equipment with other parts of the manufacturing system is becoming more common via interfacing with robots and material handling mechanisms.

A decade ago the 3D printer world was clearly defined with the hobbyist/*Rep-Rap segment*, *benchtop/office printers*, and *industrial printers*. However, the newer machines such as Markforged OnyxPro are enabling printing of composite parts in office environments. We also see a small number of machines combining laser engraving/cutting, 3D printing, and CNC machining processes. Desktop metal printers like Rapidia have become affordable but still require a sintering step and thus cannot be placed in an office environment, but a laboratory one. As the strides are made in rapid or additive manufacturing, batch size and quality issues remain but are being resolved, through faster, and larger print capabilities, and simultaneous use of different materials and colors.

1.4   Comparison of 3D Printing and Additive Manufacturing to Computer Numerically Controlled (CNC) Machining

Earlier in their development, 3D printers were expensive as a new technology, not as accurate as today's machines, did not produce parts with good surface quality, and the number processes and associated materials were limited. As mentioned earlier, early 3D printers processed paper, starch, and a limited number of thermoplastics, waxes, and epoxy resins. Thus, they fell short in ceramic, metal, and composite printing areas. Due to the developments in the last couple of decades, the 3D printing technology is able to level the playing field with CNC machining in some aspects, and the following Table 1.2 offers a general comparative analysis between the two technologies. Early processes like solid ground curing (SGC) combined deposition and machining to make rapid prototypes, similar to today's 2 in 1's or 3 in 1's with 3D printing, laser and CNC machining capabilities. An early project to design a 2 in 1 machine was successfully completed at Robert Morris University by Paul D. Badger, modifying a Denford CNC router to have 3D printing ability, involving the router machining its own 3D printer functionality's control board [24, 25].

Table 1.2 General comparison of CNC machining (excluding electro-chemical (ECM), electrical-discharge (EDM), laser, and water-jet machining) and 3D printing

Factor	CNC machining technology	3D printing technologies
Energy form	Mechanical, electro-chemical, electrical-discharge, laser, water jet pressure	Thermal, UV laser, electron beam, UV light, ultrasonic energy
Initial investment	Hobbyist and low-cost machines can be as low as a few hundred to a few thousand dollars. A used industrial CNC mill can be bought for a low price of $10–20 K. A large-scale industrial machine with added automation may cost a few $100Ks	Reliable FFF machines without enclosures can be as low as a couple of hundred dollars. FFF machines with good printing capabilities can be within $5–10 K. New metal printers cost much less than the industrially established ones and range from $150 to 200 K
Size and facility space requirements	Wide range of build-volumes are available—from benchtop to a one that you can fit a large vehicle within. Larger machines may require greater dedicated spaces	Larger and larger build volumes are being achieved. However, the upper limits are smaller than CNC technology excluding sand mold printers like Voxeljet
Geometric capabilities	Limited to parts without internal voids and cannot produce parts with intermingled segments	Can produce parts with internal voids and intermingled segments leading to very high geometric complexity
Associated materials	Polymers, ceramics, metals, composites, fabric, wood	Polymers, ceramics, metals, composites, wood, paper, living cells, food content. Certain technologies can only process specific materials—i.e. EBM can only accommodate conductors
Raw material form	Solid including sheet	Solid including sheet, liquid, powder
Cost of materials	Varies greatly	Varies greatly
Volume of production (Batch size)	Batch sizes vary from one-off to small-medium	Batch sizes have been improved with additional of material compartments, now may vary from one-off to small-medium
Dimensional accuracy and stability	Parts made are much more accurate and dimensionality stable	Have been improving and high-end machines deliver good accuracy and stability

(continued)

Table 1.2 (continued)

Factor	CNC machining technology	3D printing technologies
Surface quality	Often better than 3D printing, and excellent surface quality can be obtained with finishing and polishing operations	Have been improving, but secondary processes such as an acetone treatment can improve ABS parts printed with FDM/FFF processes
Work-holding	Clamps and vices are required	Supports and unprocessed materials cancel the need for work-holding
Secondary processing	Wide range of heat and surface treatment processes including coating have been utilized	Removal of supports, cleaning and post-curing may be required. Joining and welding of smaller parts as well as modifying the form of 3D printed parts are possible
Waste	Produces much waste (chips) compared to the 3D printing technologies. Coolants/lubricants needed are made environmentally friendly	Produces much less waste (support materials, cleaners, adhesives) compared to CNC machining. Biodegradable materials are widely used
Training	CAD/CAM and NC coding knowledge is required even with low-cost and benchtop machines	Open source printers require less training than their industrial counterparts and CNC machines
Safety and health requirements	Have been well-developed including shielding, warning lights, foot operated machine functions, and interlocks	Open volume FDM/FFF machines come with burning, electrical, and mechanical hazards. Industrial machines have better safeguarding mechanisms Requires vacuum for particular materials for fire and explosion risks. Containment of gasses and free radicals may be important for the operators of some processes like FDM/FFF Stronger ventilation and/or filtration of air is also needed for SLA machines or similar
Ancillary equipment	Jigs, fixtures, and cutting tools are required along with automated tool storage and changing, chip removal, and lubricant/coolant management systems	Cleaning and support removal, debinding, post-curing, and sintering equipment may be needed

Review Questions

1. Define each industrial revolution (Industry 1.0 through 4.0) concisely.
2. What are the four cornerstones of the CIM concept defined by Mikell P. Groover? Explain each one in detail.
3. What is smart manufacturing? How does it relate to CIM and Industry 4.0?
4. Define digital manufacturing.
5. Define rapid prototyping, tooling, and manufacturing concepts in one sentence each.
6. Draw the workflow for digital manufacturing, through two different pathways: 3D printing/AM or CNC machining.
7. What are the three different ways CAE is involved in design and manufacturing of products or machines? Elaborate each concisely.
8. Describe the role of digital manufacturing in product design and development.
9. What is the oldest known technological development relevant to 3D printing?
10. Compare CNC machining to 3D printing in 5 different aspects.

Research Questions

1. What is Industry 5.0? Write a paragraph about it.
2. Conduct a literature review to find a case study on digital twin applications in manufacturing, and summarize it in a single powerpoint slide.
3. Conduct a literature review to find a case study on digital thread applications in manufacturing, and summarize it in a single powerpoint slide.

Discussion Questions

1. After studying the concept of 4D printing, discuss a potential application of 4D printing or develop an alternative 4D printing process or material concept.
2. Company X is a small manufacturer. It is trying to develop a new solution as a supplier working under tight deadlines. The 3D printer owned by the company has a small build volume, about half the size needed and the company has no other choice but to use their own printer to print this ABS part. How can the company build the prototype using this printer?

References

1. Howard, E. (2018). The Evolution of the Industrial Ages: Industry 1.0 to 4.0, posted in Industry 4.0, Scheduling, Simulation blog, located at: https://www.simio.com/blog/2018/09/05/evolution-industry-ages-industry-1-0-4-0/, accessed May 31, 2020.

2. Crandall, R. E. (2017). Industry 1.0 to 4.0: The Evolution of Smart Factories, SCM Now Magazine, September/October 2017, located at: https://www.apics.org/apics-for-individuals/apics-magazine-home/magazine-detail-page/2017/09/20/industry-1.0-to-4.0-the-evolution-of-smart-factories, accessed May 31, 2020.
3. Groover, M. P. (2003). *Automation, Production Systems, and Computer-Integrated Manufacturing 2nd Edition*, Upper Saddle River, NJ: Prentice Hall. 8
4. Leiva, C. (2015). On the Journey to a Smart Manufacturing Revolution, located at: https://www.industryweek.com/technology-and-iiot/systems-integration/article/21967056/on-the-journey-to-a-smart-manufacturing-revolution?page=2, accessed May 31, 2020.
5. Erdem, E. (2017). AutoDesk Moldflow as a Tool for Promoting Engaged Student Learning, American Society for Engineering Education (ASEE) Annual Conference and Exposition, Columbus, OH.
6. PAM-Crash Software, ESI Group, located at: https://www.esi-group.com/pam-crash, accessed May 31, 2020.
7. DeCesaris, D., Bonsano, F. (2018). Thingworx ANSYS—Value of IOT and Simulation, ANSYS Innovation Conference, Bologna, Italy.
8. MacDonald, C., Dion, B., Davoudabadi, M., Creating a Digital Twin for a Pump, ANSYS Corporate Resource Library, located at: https://www.ansys.com/-/media/ansys/corporate/resourcelibrary/article/creating-a-digital-twin-for-a-pump-aa-v11-i1.pdf, accessed November 4, 2019.
9. Levia, C. (2015). What is the digital thread? iBaseT, located at: https://www.ibaset.com/blog/what-is-the digital-thread, accessed May 31, 2020.
10. Ntop Platform, Ntopology, located at: https://ntopology.com/ntop-platform/, accessed May 31, 2020.
11. Ulrich, K.T., Eppinger, S.D. (2012). *Product Design and Development 5th Edition*. New York, NY: McGraw-Hill Irwin. 2.3.15.16
12. Greguric, L. (2018). History of 3D printing—When was 3D printing invented? all3dp, located at: https://all3dp.com/2/history-of-3d-printing-when-was-3d-printing-invented/, accessed May 31, 2020.
13. Goldberg, D. (2018), History of 3D Printing: It is Older Than You are (That is If You're under 30), Autodesk Redshift, located at: https://autodesk.com/redshift/history-of-3d-printing, accessed May 31, 2020.
14. Rep-Rap Wiki, RepRap Project, located at: https://reprap.org/wiki/RepRap, accessed May 31, 2020.
15. Kickstarter, Kickstarter, located at: https://www.kickstarter.com/about, accessed May 31, 2020.
16. Voxeljet Sand Printers, Voxeljet, located at: https://www.voxeljet.com/3d-drucksysteme/vx4000/, accessed May 31, 2020.
17. ExOne Sand Printers, ExOne Company, located at: https://www.exone.com/en-US/3D-printing-systems/sand-3d-printers, accessed May 31, 2020.
18. Sirinterlikci, A., Czajkiewicz, Z., Doswell, J., Behanna, N. (2009). Direct and Indirect Rapid Tooling, Rapid/3D Scanning Conference, Chicago, IL.
19. Sirinterlikci, A., Uslu, O., Behanna, N., Tiryakioglu, M. (2010), International Journal of Modern Engineering, 10 (2), pp-NA.
20. GE Aviation acquires Morris Technologies and Rapid Quality Manufacturing, Press Center, located at: https://www.geaviation.com/press-release/other-news-information/ge-aviation-acquires-morris-technologies-and-rapid-quality, accessed November 4, 2019.
21. Kellner, T. (2016). All the 3D Print That's Fit to Pitt: New Additive Technology Center Opens Near Steel Town, located at: https://www.ge.com/reports/all-the-print-that-fits-to-pitt-new-additive-technology-center-opens-near-steel-town/, accessed November 4, 2019.

22. GE makes significant progress with investments in additive equipment companies, GE Additive Press Release, located at: https://www.ge.com/additive/press-releases/ge-makes-significant-pro gress-investments-additive-equipment-companies, accessed November 4, 2019.
23. Concept laser, GE Additive, located at: https://www.ge.com/additive/who-we-are/concept-laser, accessed November 4, 2019.
24. Sirinterlikci, A., Badger, P.D., Mura, C., Kronz, F., Tabassum, R., Swink, I. (2013) Designing and Building Open Source Rapid Prototyping Machines, Rapid Conference, Pittsburgh, PA.
25. Badger, P.D., Sirinterlikci, A., Yarmeak, G., (2016), Conversion of a 3-Axis Commercial Milling Machine into a 3D Printer, Journal of Engineering Technology, 33(2),40–48.

Preparation for 3D Printing and CNC Machining

2

Delivery of an accurate prototype or an end-user part relies on two critical steps along the digital manufacturing workflow—computer-aided design (CAD) assuring an accurate representation of the intended design and computer-aided manufacturing (CAM) for converting that design into an accurate part. This chapter covers the CAD fundamentals relevant to digital manufacturing including the CAD architecture and different forms of modeling such as wireframe, surface, solid, and their applications. CAD's role in CNC machining and 3D printing are the topics of the Chaps. 3 and 4 respectively.

2.1 Computer-Aided Design (CAD) Subjects Relevant to 3D Printing and CNC Machining

Before reviewing the CAD subjects relevant to 3D printing and CNC machining, one needs to fully understand the CAD architecture and its different forms of modeling. A CAD system has three major parts to it: the hardware including the computer and its input/output (I/O) devices, the operating system, and the CAD application software as illustrated in Fig. 2.1 [1]. The application software runs on the operating system of the computer and is the main software component tying the user requests from the user interface with the graphics utility and the CAD model database to accomplish them. The graphics utility performs functions like coordinate transformations for moving the objects including translation, rotation, and zooming through scaling, windowing and display control. Device drivers allow communication between the graphics utility, the input and output devices, and the user interface.

© The Author(s), under exclusive license to Springer Nature Switzerland AG 2023
A. Sirinterlikci and Y. Ertekin, *A Comprehensive Approach to Digital Manufacturing*,
Synthesis Lectures on Mechanical Engineering,
https://doi.org/10.1007/978-3-031-25354-6_2

Fig. 2.1 CAD system
architecture [1]

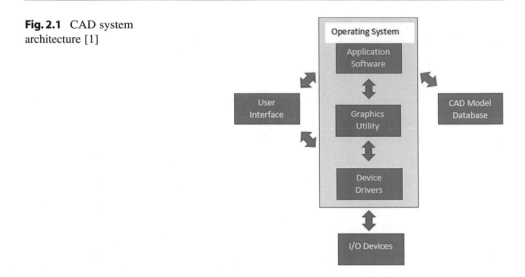

2.1.1   Wireframe Modeling Including Analytical and Free-Form Curves

Wireframe CAD models are composed of points, lines, and curves. They require less computing time and memory space, but cannot provide enough information about a 3D object's surface details including discontinuities [2]. Figure 2.2 illustrates ambiguity caused by a 3D wireframe model. The hollow prismatic part design can be interpreted by a CAD operator a few different ways including the two shown below by shading those non-existent surfaces.

Not only were early CAD software tools able to generate 3D wireframe models for engineers to see their designs including assemblies, but also allowed numerical control (NC) tool path generation, robot programming, integrated and printed circuit layout development as well as structural analysis [1]. Today's solid modelers still utilize wireframe

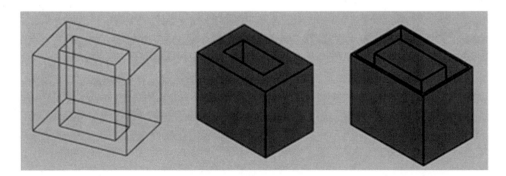

Fig. 2.2 Ambiguity caused by wireframe models

Fig. 2.3 A 2D Mastercam wireframe model (in green) being utilized to generate an NC tool path (blue/yellow) for machining a 3D part with constant cross-section through the z-axis

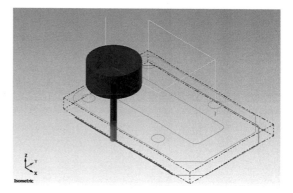

capability under their display controls enabling users to see the details of an intricate part, and a CAM software tool such as Mastercam still employs wireframe modeling in its tool path generation. Figure 2.3 illustrates a cutting tool following the selected wireframe features such as lines and curves to backplot (playback) an NC tool path. The toolpath is made from two components: blue color representing path with a prescribed feed rate while yellow lines are for rapid traverse.

2.1.1.1   Curves

There are two types of curves: *analytical and free-form.* Analytical curves are defined by known equations and are exact [1]. On the contrary, control points drive the free form curves.

Analytical Curves

There are two basic ways to define a curve: a *nonparametric way, as a function of positional variables (x, y, and z),* and a *parametric way, as a function of another parameter (such as θ)* [1]. Nonparametric curves can be represented *implicitly* (Eq. 2.1) or *explicitly* where one variable is prescribed as a function of the others (Eq. 2.2).

$$f(x, y, z) = 0 \tag{2.1}$$

$$z = f(x, y) \tag{2.2}$$

Analytical curves are commonly represented in explicit form since it is hard to plot them in implicit form, given in Eq. 2.3 for a circle:

$$(x - x_0)^2 + (y - y_0)^2 - r^2 = 0 \tag{2.3}$$

where x_0, y_0 are the coordinates of the center of the circle and r is its radius.

Equation 2.4 is the explicit form of the Eq. 2.3 for the circle after the conversion process is completed. Each x value will yield two y values for the circle as the x values fall within the range of $x_0 - r \leq x \leq x_0 + r$.

$$y = y_0 \mp \sqrt{r - (x - x_o)^2} \tag{2.4}$$

The third form of a circle is the parametric form, given below in Eqs. 2.5 and 2.6, and is based on the parameter θ being applied to the radius, r:

$$x = r\cos\theta + x_0 \tag{2.5}$$

$$y = r\cos\theta + y_0 \tag{2.6}$$

where θ lies within the range of $0 \leq \theta \leq 2\pi$.

Major analytical curves are obtained through sectioning a cone including a circle, ellipse, parabola, and hyperbola. These curves are given in the implicit form as shown in Table 2.1.

Freeform Curves

Free-form curves can be approximated by a polynomial equation and if the shape is too complex to be handled by a single equation, it can be broken into multiple segments. To define a free-form curve, it is necessary to determine the coefficients of the polynomial equation. However, that step is not enough to determine the shape of the curve, an additional step needs to be taken, which is explained in this section [1].

A third-degree polynomial, a cubic curve, will be satisfactory for representing most curve applications. A general cubic parametric curve r (t) is given below in Eqs. 2.10 and 2.11 where $0 \leq t \leq 1$. Its components (x, y, and z) are also prescribed in parametric form as a function of t in Eqs. 2.7–2.9 respectively.

$$x(t) = a_x t^3 + b_x t^2 + c_x t + d \tag{2.7}$$

$$y(t) = a_y t^3 + b_y t^2 + c_y t + d \tag{2.8}$$

$$z(t) = a_z t^3 + b_z t^2 + c_z t + d \tag{2.9}$$

$$r(t) = a t^3 + b t^2 + c t + d \tag{2.10}$$

or

$$r(t) = CT^t \tag{2.11}$$

where C and T are defined in matrix space

Table 2.1 Analytical curve definitions and generation

Analytical curve type	Implicit formula	Analytical geometry definition
Circle	$(x - x_0)^2 + (y - y_0)^2 - r^2 = 0$	
Ellipse	$\frac{x^2}{a^2} + \frac{y^2}{b^2} - 1 = 0$	
Hyperbola	$\frac{x^2}{a^2} - \frac{y^2}{b^2} - 1 = 0$	
Parabola	$y^2 - 4ax = 0$	

$$r(t) = \begin{bmatrix} x(t) \\ y(t) \\ z(t) \end{bmatrix} \tag{2.12}$$

$$C = \begin{bmatrix} a & b & c & d \end{bmatrix} = \begin{bmatrix} a_x & b_x & c_x & d_x \\ a_y & b_y & c_y & d_y \\ a_z & b_z & c_z & d_z \end{bmatrix} \tag{2.13}$$

$$T = [t^3 \ t^2 \ t1] \tag{2.14}$$

The tangent of the curve, r(t), can determined by taking a derivative of it:

$$\dot{r(t)} = \frac{dr(t)}{dt} = 3at^2 + 2bt + c \tag{2.15}$$

Fig. 2.4 Ferguson's curve

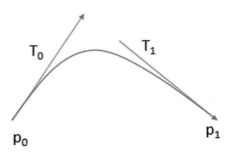

In order to define a long curve with multiple curve segments, one must be able to control the end-points of the curve and the tangent at the end points. By setting the end point of the first segment as the beginning point of the second segment allows generation of a continuous curve without a break, a condition called the *zero order (1st degree) continuity*. If the tangent at the end-point of the first segment is in the same direction as the tangent of the beginning point of the second segment, the combined curve will *be smoother, a condition called the first order (2nd degree) continuity.*

Multiple special curve definitions were developed. The following section will cover these free-form curve definitions including *Ferguson's, Bezier's, B-spline and NURB* curves.

Ferguson's Curve
Ferguson's curve, r(t), is defined by two endpoints (p_0, p_1) and two tangent vectors (T_0, T_1) at the end points [1] and characterized mathematically below via Eqs. 2.16–2.19 (Fig. 2.4):

$$r(0) = p_0 \tag{2.16}$$

$$r(1) = p_1 \tag{2.17}$$

$$\dot{r}(0) = T_0 \tag{2.18}$$

$$\dot{r}(1) = T_1 \tag{2.19}$$

By substituting t = 0 and t = 1 into Eqs. 2.10 and 2.15, the following set of equations (Eqs. 2.20–2.23) are obtained, also employing help from relationships given above in Eqs. 2.16–2.19:

$$p_0 = r(0) = d \tag{2.20}$$

Fig. 2.5 Multiple curve segments used for generating a complex curve in Solidworks

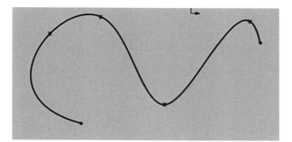

$$p_1 = r(1) = a + b + c + d \qquad (2.21)$$

$$T_0 = \dot{r}(0) = C \qquad (2.22)$$

$$T_1 = \dot{r}(1) = 3a + 2b + c \qquad (2.23)$$

Therefore, the following solution is achieved,

$$a = 2p_0 - 2p_1 + T_0 + T_1$$

$$b = -3p_0 + 3p_1 - 2T_0 - T_1$$

$$c = T_0$$

$$d = p_0$$

allowing Eq. 2.10 to be rearranged with the known coefficient values of a, b, c, d.

$$r(t) = [p_0 \ p_1 \ T_0 \ T_1] \begin{bmatrix} 2 & -3 & 0 & 1 \\ -2 & 3 & 0 & 0 \\ 1 & -2 & 1 & 0 \\ 1 & -1 & 0 & 1 \end{bmatrix} \begin{bmatrix} t^3 \\ t^2 \\ t \\ 1 \end{bmatrix}$$

where $0 \leq t \leq 1$.

Multiple Ferguson's curve segments can be utilized to make up a complex curve that is smooth and continuous, similar to the complex curve given below in Fig. 2.5. However, the tangents for the curve endings do not relate to the overall form of the complex curve.

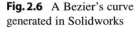
Fig. 2.6 A Bezier's curve generated in Solidworks

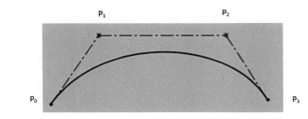

Bezier's Curve

Bezier's curve was developed in the early 1970s by Dr. Bezier who was a staff member at the French Renault automotive company [1]. It is still commonly used in CAD packages. It can be defined as a lower- or higher-order curve. The cubic version of the Bezier's curve is defined by four control points or vertices (p_0, p_1, p_2, p_3) as illustrated in Fig. 2.6. The control points shown in the figure below develop a fraction of a polygon and are effective to shape the curve. The relationships between the four control points and a Bezier curve segment are given below in Eqs. 2.24–2.27.

$$r(0) = p_0 \tag{2.24}$$

$$r(1) = p_3 \tag{2.25}$$

$$\dot{r(0)} = 3(p1 - p0) \tag{2.26}$$

$$\dot{r(1)} = 3(p3 - p2) \tag{2.27}$$

By substituting $t = 0$ and $t = 1$ into Eqs. 2.10 and 2.15 the following set of equations are obtained (Eqs. 2.28–2.31), also employing help from the relationships given above in Eqs. 2.24–2.27:

$$p_0 = r(0) = d \tag{2.28}$$

$$p_3 = r(1) = a + b + c + d \tag{2.29}$$

$$3(p1 - p0) = \dot{r(0)} = c \tag{2.30}$$

$$3(p3 - p2) = \dot{r(1)} = 3a + 2b + c \tag{2.31}$$

Therefore, the following solution is achieved,

$$a = p_3 - 3p_2 + 3p_1 - p_0$$

$$b = 3p_2 - 6p_1 + 3p_0$$

$$c = 3p_1 - 3p_0$$

$$d = p_0$$

allowing Eq. 2.10 to be rearranged with the known coefficients, a, b, c, d,

$$r(t) = \begin{bmatrix} p_3 & p_2 & p_1 & p_0 \end{bmatrix} \begin{bmatrix} 1 & 0 & 0 & 0 \\ -3 & 3 & 0 & 0 \\ 3 & -6 & 3 & 0 \\ -1 & 3 & -3 & 1 \end{bmatrix} \begin{bmatrix} t^3 \\ t^2 \\ t \\ 1 \end{bmatrix}$$

where $0 \leq t \leq 1$.

Cubic Bernstein polynomials (also called blending functions) can also be used in expressing the r(t) function. This chapter will not cover the Bernstein polynomials [1]. However, their impact on understanding the shape of the Bezier's curves are too important to neglect. Changing any of the control points will result in a change of the entire curve with one exception, the beginning and end points have no impact on the shape of the curve. In order to achieve zero–order continuity (1st degree), the first point of the second curve segment must be the same as the fourth point of the first curve segment. For the first-order continuity (2nd degree), the last two points of the first curve segment and the first two points of the second curve segment must be collinear.

B-splines
Four different variations of Basis- or B-Splines exist excluding their orders or degrees:

- A *uniform B-Spline (called Cardinal B-Spline)* is where *the distance between the control (through) points (knot spacing or span) is equal* throughout the curve (Fig. 2.7).
- A *non-uniform B-Spline* is where the knot spacing varies throughout the curve (Fig. 2.8).
- A *rational B-Spline is a uniform B-Spline where each control point has a weight* (Fig. 2.7).
- A NURB or the *non-uniform rational B-Spline* is where we have a *non-uniform B-Spline that is also rational* (Fig. 2.8).

When a Bezier's curve is generalized to degree-n, n + 1 control points are blended together, making the case for a B-Spline. Thus, a non-uniform B-spline is a general case of a Bezier's curve. A B-spline is defined with a *Cox-deBoor recursive function* given

Fig. 2.7 A uniform rational
B-Spline compared to a
uniform
B-Spline—The rational spline
has only its second control
point with an added weight
(2.0) pulling the curve towards
itself and away from the
uniform spline

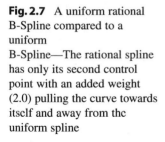

Fig. 2.8 The lighter colored
curve is a NURB (with even
numbered control points
having 2 times greater weights
than the odd numbered ones)
while the other is a
non-uniform B-Spline with no
weights involved

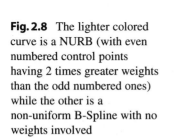

below (Eq. 2.32) [1]:

$$r(t) = \sum_{j=0}^{L} N_j^n(t) p_j \qquad (2.32)$$

where

$$N_j^n(t) = \frac{t - t_i}{t_{i+n-1} - t_i} N_i^{n-1}(t) + \frac{t_{i+n} - t}{t_{i+n} - t_{i+1}} N_{i+1}^{n-1}(t) \qquad (2.33)$$

$$N_i^1(t) = \{1, t \in [t_i, t_{i+1}] \ and \ t_i < t_{i+1} 0, \ otherwise$$

$$L = \text{number of control points}$$

$$n = \text{degree of the curve.}$$

Employing a difference operator (∇) algebraic equations above can be simplified:

$$\nabla_i = t_{i+1} - t_i \qquad (2.34)$$

$$\nabla_i^k = \nabla_i + \cdots + \nabla_{i+k-1} t_{i+k} - t_i \qquad (2.35)$$

Yielding the following equation,

$$N_j^n(t) = \frac{t - t_i}{\nabla_i^{n-1}} N_i^{n-1}(t) + \frac{t_{i+n} - t}{\nabla_{i+1}^{n-1}} N_{i+1}^{n-1}(t) \tag{2.36}$$

The difference operator, ∇, describes the knot spacing, the distance between the neighboring control points as mentioned earlier. A non-uniform B-spline will have non-constant knot spacing while a uniform B-spline will have constant ones. The cubic uniform B-spline blending functions indicate that control points will only have a local effect on the curve. The same conclusion can be drawn for a long curve represented by a large number of control points. As mentioned earlier, the B-spline is a general case of a Bezier's curve. This also makes Bezier's curve a special case of a uniform B-spline (where the knot spacing is not varying).

A vector [x y z] represents the coordinates of a point P in 3D space. A homogeneous vector is four dimensional [x y z h] and to project it to a plane h = 1, the vector takes a rational form [x/h y/h z/h 1] by normalization [1]. Then, h is considered as weight. If $h \neq 0$, the two vectors are identical in 3D space. In a similar fashion, the vector [x y z 1] is the same as [xw yw zw w]. Thus, utilizing the vector [$x_i w_i$ $y_i w_i$ $z_i w_i$ w_i] for pi in the definition of a Bezier's curve in polynomial form $B_i(t)$ (Eq. 2.37) and normalizing the resultant curve yields a rational Bezier's curve as follows:

$$r(t) \frac{\sum B_i(t) w_i p_i}{\sum B_i(t) w_i} \tag{2.37}$$

The shape of the resulting curve can be changed by changing the weights, varying from a straight line to the characteristic polygon when w_i is positive [1] (Figs. 2.7 and 2.8)

In Practice

Solidworks 2021 style spline functionality is capable of generating B-Splines of 3°, 5°, or 7° in addition to a Bezier's curve [3]. Conversion from a spline to a style spline is possible and ends with a B-Spline of 3°. B-Splines with higher degrees require a larger number of minimum control points (i.e. 3° requires 4, 5° requires 6 and 7° requires 8). Please follow the instructions below to draw three curves similar to the ones given in Solidworks using the same control points for all three types while changing the style spline types to study the changes in the curves

- Tools > Sketch Entities > Style Spline.
- In the Insert Style Spline Property Manager, select the B-Spline 3 curve from options.
- Select the control points in the sketch plane.

- Repeat the process using the same control points but alter the spline type to B-Spline 5° and B-Spline 7° while superimposing the curves.

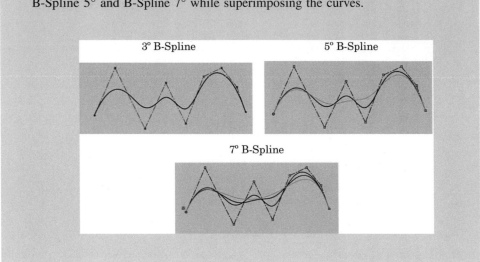

NURBs

The *NURB* is a *non-uniform rational B-spline*. It carries the features of all of the preceding curve types, making them each a special type of a NURB. A NURB has a non-uniformly distributed (knot spacing) set of control points with weights (Fig. 2.8). NURBs are flexible and can fit to complex shapes, but they are also much more complex to compute and use. NURBs are also available in various CAD packages.

2.1.2 Surface Modeling

Surface modeling has many applications including aesthetic surfaces such as vehicles and a variety of craft bodies like aircraft or sea craft and consumer product outer-forms, which also serve a function (*class-A surfaces or Strak*) and technical surfaces such as mold and die cavity, turbine blades, impeller and propeller designs. Class-A modeling is very common in automotive manufacturing and describes the challenge of combining the need for [4]:

- Continuity across the surface patches and having perfect highlight reflections
- Engineering requirements—flanges and fit
- Production requirements—modeling tolerances

Fig. 2.9 Mastercam's simple surface model creation feature includes pre-defined primitives such as a cylinder, block, cone, sphere, and torus

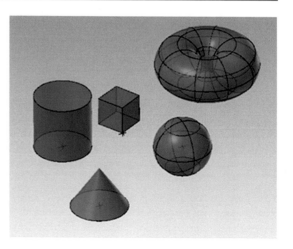

Fig. 2.10 Free-form surfaces [5]

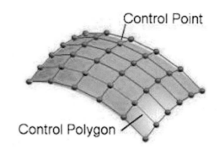

Just like the curves, *surfaces* can be classified as *analytical and free-form* [1]. Analytical surfaces such as *circular, cylindrical, spherical, conical, and tori* are defined by known equations (Fig. 2.9). On the other hand, *the free-from (sculptured) surfaces* are based on *free-form curves including Bezier's, B-Splines, and NURBs* (Fig. 2.10).

Multiple methods are utilized in making of surface models [1]:

- Use of a profile curve (labeled as an *across contour* in Mastercam) which can be analytical or free-form and sweeping it along a different curve (labeled as along contour in Mastercam) or extruding it along a line (Fig. 2.11), revolving it about an axis, or employing multiple curves for blending them or generating a mesh between them by lofting through.
- Direct generation of free-form surfaces employing control points and characteristic (control) polygons as shown in Fig. 2.10.
- Creating new surfaces from existing surfaces via off-sets, bridging, and blending.

Fig. 2.11 A tabulated cylinder
obtained in Solidworks

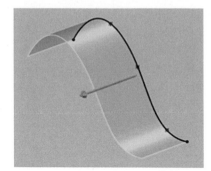

Fig. 2.12 A cylindrical
surface obtained by extrusion
in Solidworks

In a general case, *a tabulated cylinder* r(t, s) is created employing a curve r(t) and extruding or sweeping along a vector N (or simply extruding) as shown in Fig. 2.11 based on Eq. 2.38 [1]. When the r(t) is a circle, the operation yields a cylindrical surface (Fig. 2.12) whereas the r(t) is a line, resulting in a rectangular surface.

$$r(t, s) = r(t) + sN \quad \text{where } 0 \leq t \text{ and } s \leq 1 \tag{2.38}$$

A curve r(t) can be rotated (in the amount of an angle) θ around an axis to obtain multiple surfaces (Eq. 2.39), in the case of a line base shape such as a rectangular wireframe yielding a cylindrical surface or a triangle yielding a conical surface [1].

$$r(t, \theta) = [x(t)\cos\cos\theta\, x(t)x\sin\sin\theta\, z(t)] \quad \text{where } r(t) = [x(t)\, 0\, z(t)] \tag{2.39}$$

Blending of two curves ($r_0(t)$ and $r_1(t)$) results in a ruled surface (Fig. 2.13).

$$r(t, s) = r0(t) + s[r1(t) - r0(t)] \quad \text{where } 0 \leq t \text{ and } s \leq 1 \tag{2.40}$$

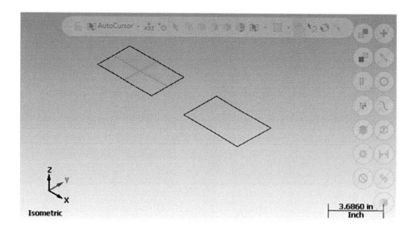

Fig. 2.13 Ruled surface in Mastercam—using blending of short and long edges of a rectangle marked by green lines

Free-Form Surfaces

Free-form surfaces were defined in the Sect. 2.1.2 above including two different ways of generation. However, engineers need to rely on the CAD tools available to them. The lead author often uses the surface creating features of Mastercam software for CNC work. Thus, multiple surface generation examples and associated surface model types were included below in the In-Practice Section. Solidworks has a set of surface features while Autodesk has a special tool called Alias for Class A modeling. Another software which may be beneficial for its surface modeling and free-form strength is Rhinoceros 3D.

2.1.3   Solid Modeling

A solid model captures a complete geometry of a 3D object, allowing its volume and mass to be calculated with a material assignment (via its density) as it differentiates its interior and exterior. A solid model can hide some of its surfaces and lines using its other surfaces, shading features, and can also be converted into a wireframe in display mode.

There have been multiple methods developed for generation of the solid models: *primitive instancing, spatial occupancy enumeration, cell decomposition, constructive solid geometry, boundary representation, and sweeping* [1].

- Primitive instancing is based on utilizing pre-defined geometries (Fig. 2.14) and prescribing their dimensions for generation of family of parts. It is similar to the Group Technology concept utilized in cellular manufacturing [6, 7]. Even though these types

Fig. 2.14 Part family A and B
(with a function of D_1, L_1, D_2,
and L_2)

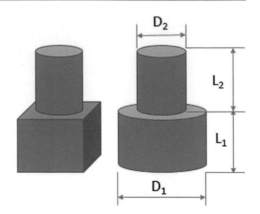

of models are concise, it is difficult to conduct a geometric operation on them. These models need to be converted to another type of models such as a B-Rep one before any operation can be performed on them [1].

- Spatial occupancy enumeration (SOE) uses very small identical spatial cells to make up 3D geometry (Fig. 2.15). To be able to accurately define any geometry within 0.001 in, it has to use cells with a volume of 10^{-9} in. Each cell location is defined in 3D space with its x,y, and z coordinates, however generating a small volume requires such large memory making this method not practical [1].

In Practice

Complete the following Mastercam surface creation exercises based on the wireframe models given (in pink). You can also use another CAD package if you do not have access to Mastercam. Generate:

- A draft surface with no angle, and add a draft angle of your choice

Fig. 2.15 Spatial occupancy
enumeration

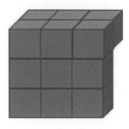

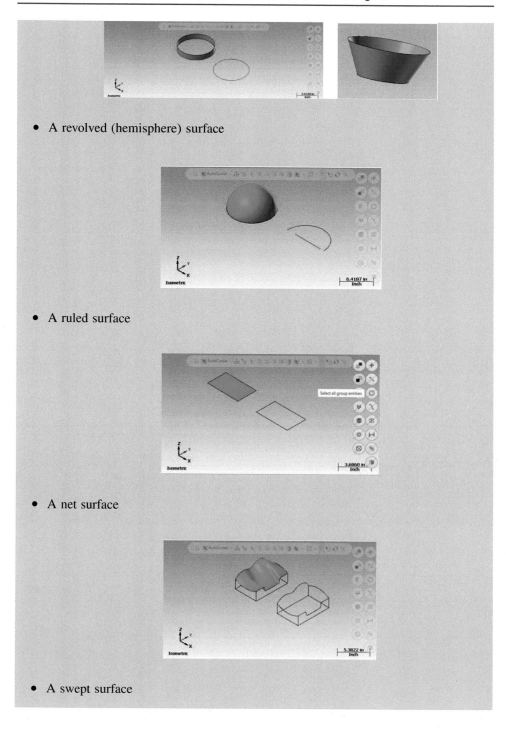

- A revolved (hemisphere) surface

- A ruled surface

- A net surface

- A swept surface

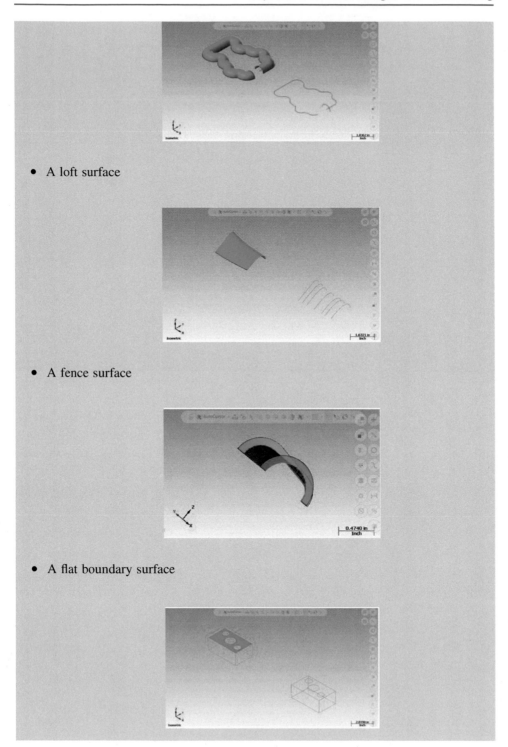

- A loft surface

- A fence surface

- A flat boundary surface

- A surface fillet

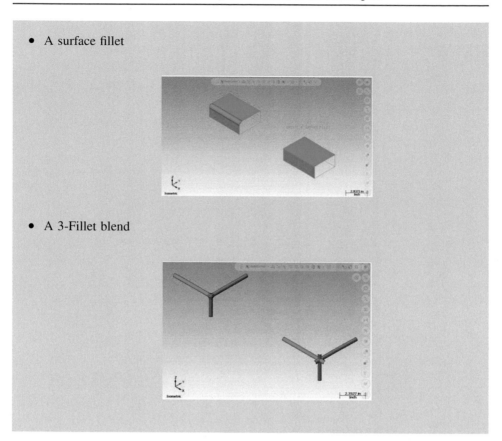

- A 3-Fillet blend

- Cell decomposition (Fig. 2.16) method is a special form of the SOE method mentioned above, with a difference of variable cells requiring lesser computing resources. The solid is broken down into solid cells, whose interiors are pairwise disjoint. These cells are defined with their dimensions and locations (x, y, and z coordinates). Boolean operations are used to further manipulate the geometry of the objects [1].

Fig. 2.16 Cell decomposition

Fig. 2.17 Constructive solid
geometry, employing
difference (-) and union (U)
Boolean operations as well as
three primitives, A, B, and C

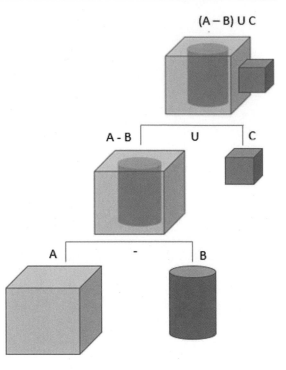

- Constructive solid geometry (CSG) is an effective system for creating 3D geometry, employing primitives (such as prisms, spheres, cones etc.) and Boolean operations (such as union, difference, and intersection). A CSG model is represented by a model tree (Fig. 2.17) [1].

- Boundary representation (B-Rep) method is based on using the 3D geometry's boundaries (Fig. 2.18). Most boundary models are driven by evaluating a CSG model of objects. The B-Rep model of an 3D geometry is only made from bounded faces with no loose edges or faces, due to the manifold modelers using the Euler formula. Each edge is bounded by two vertices and is adjacent to two faces. Each vertex belongs to one edge. Non-manifold modelers allow additional faces and edges to exists in a solid model [1].

- Sweeping is accomplished by use of revolution and translation. If half of a rectangle is revolved for 360°, it generates a cylinder in the solid modeling mode of CAD programs. If the same rectangle is extruded within a certain distance, it produces a prism, which can also be done using a linear guide (along contour) and the sweeping functions [1].

Fig. 2.18 Boundary
representation of a prismatic
block (vertices and edges of
boundary surfaces are not
shown in the figure, but need
to be defined)

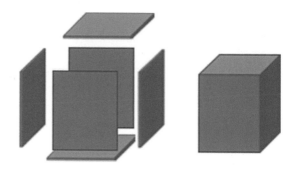

Over the years CSG and B-Rep methods have dominated the solid model software tool
design as the other methods had obvious shortcomings or issues in resource requirements
including processing speed.

> **In Practice**
> Use the Boolean operations (after designing or finding a pattern) available in Solid-
> works to design a mold insert for an injection molding set-up for both of its cover
> and ejector halves by following Solidworks help tools or tutorials from the Internet.
> Make sure that the tooling has an appropriate cavity for real-life use including the
> feeding system (the runner, the gate, and the sprue channel)
>   Hint: Chap. 7 includes an injection molding tool making case study

Review Questions

1. Define the basic structure of CAD architecture using a diagram and explain functions
   of its main components and interactions between them.
2. Define a wireframe CAD model including its pros and cons and application areas.
3. What are the differences between an analytical curve and a free-form curve?
4. List at least five types of analytical curves.
5. Represent a circle with center (5, 5) and a radius of 5 in using implicit and explicit
   forms along with parametric form.
6. List at least five types of free-from curves.
7. Find the tangent line of the curve $y = 10 + 5x + x^2$ at $x = 2$
8. Find the tangent line of a parametric curve given below at $x = 2$

$$r(t) = CT^t$$

$$C = \begin{bmatrix} a\ b\ c\ d \end{bmatrix} = \begin{bmatrix} a_x\ b_x\ c_x\ d_x \\ a_y\ b_y\ c_y\ d_y \\ a_z\ b_z\ c_y\ d_z \end{bmatrix} = \begin{bmatrix} 1\ 0\ \ \ 4\ \ \ 5 \\ 2\ 3\ -2\ 8 \\ 1\ -1\ 3\ \ \ 9 \end{bmatrix} \quad T = \begin{bmatrix} t^3 t^2 t 1 \end{bmatrix}$$

9. Define the Ferguson's curve.
10. Find a numerical example of a Ferguson's curve and explain the outcome curve based on the inputs of the example.
11. Define the Bezier's curve.
12. A Bezier's curve is defined by four control points (3, 0, 2), (4, 0, 5), (8, 0, 5) and (10, 0, 2). Find the equation of the curve in matrix form.
13. Define the B-Spline and its relation to the Bezier's curve
14. Define uniform, non-uniform, and rational B-Splines.
15. Define the NURB curve and its relation to the B-Spline.
16. Define a surface model including its pros and cons and application areas.
17. What are the differences between an analytical and a free-form surface?
18. List at least five types of analytical surfaces.
19. List at least five types of free-from surfaces.
20. Define a solid model including its pros and cons and application areas compare to the other CAD geometry.
21. What are the major differences between a wireframe and solid model?
22. Define methods used in solid model generation utilizing examples including primitive instancing, spatial occupancy enumeration, cell decomposition, construction solid geometry, and boundary representation.

Research Questions

1. How can you generate an equation driven curve in Solidworks?
2. Research the Classification and Coding (C&C) system and Group Technology (GT). How can it relate to Primitive Instancing modeling?
3. A solid model can be emptied out using a shell function in a CAD program like Solidworks to obtain a thin walled shell model. What type of CAD or CAE applications use shell models?

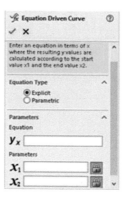

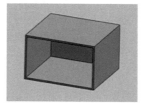

4. Compare and contrast the surface features of Mastercam, Creo, and Solidworks. Which one does seem to be the best one for surface modeling? Why?

Discussion Questions and Problems

1. Develop a 3D wireframe model with low perceived ambiguity in Solidworks or Mastercam.
2. Insert a 2D drawing or photograph of a land vehicle or sea or aircraft of your choice into a CAD program, preferably in Solidworks, and try to fit any of the curve types available to the outline of the image.
3. What surface functionalities can be employed in Mastercam to accomplish the following faucet design shown in the figure?

4. The ruled surface generated in Mastercam of the following model with two curves will be similar to the lofted one. What is the difference between ruled and lofted surface generation (sweep)?

5. The race car model shown below is generated using what type of surface type in Mastercam. Based on the same principle, develop your own race car body model by sketching multiple curves and use them to generate the body model.

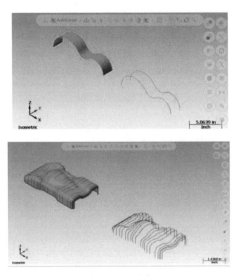

6. Following wireframe model is being used for generating two different models (a bottle and a candleholder) with revolving the profile curve about two different rotational axes. Please imitate this process using Mastercam or a similar software with surface features.

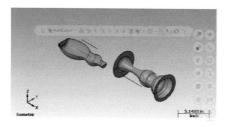

7. Develop the CSG model tree of the part given below (Hint: Assume that it is a shell model with no bottom face).
8. Generate the B-rep model of the part given for question 7?

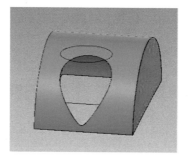

References

1. Chang, T.-C., Wysk, R.A., Wang, H.-P., (1998). *Computer-Aided Manufacturing 2nd Edition.* Upper Saddle River, NJ: Prentice Hall.
2. Amirouche, F.M. L. (1993). *Computer-Aided Manufacturing*, Upper Saddle River, NJ: Prentice Hall.
3. Splines -2021, SOLIDWORKS Help, located at: https://help.solidworks.com/2021/english/SolidWorks/sldworks/c_splines.htm, accessed September 22, 2022.
4. Understanding Class A Modeling, located at: https://knowledge.autodesk.com/Alias-Tutorials/Files, accessed September 22, 2022.
5. Class A surface—Wikipedia, located at: https://en.wikipedia.org/wiki/Class_A_surface/, accessed September 22, 2022.
6. Groover, M. P. (2003). *Automation, Production Systems, and Computer-Integrated Manufacturing 2nd Edition*, Upper Saddle River, NJ: Prentice Hall.
7. Requicha, A. A. G. (1980). Representations of Rigid Solid: Theory, Methods, and System, Computing Surveys, 12 (4), 437–464.

Computer-Aided Manufacturing (CAM)

3

3.1 Historical Background of Numerical Control (NC) and CNCs

In the early 1800s the basic concept of control was implemented in the weaving industry. NC began a few years after World War II with the requirement for stronger and lighter jet airplane structures. Two choices existed to produce these structures, e.g., wing, fuselage, bulkhead, riveting together many parts or machining a wing or fuselage from a solid piece. The one-piece structure was generally stronger and weighed less. The US Air Force started making these parts using *tracer type milling machines*. The amount of machining improved with increased speed of new aircraft design. Tracer type milling was pushed to the limit and could not cope with the amount of machining and complexity. Mr. John T. Parsons, resident of Traverse City, MI, conceived the principle of NC machine tool movement in1940. Mr. Parsons demonstrated his idea of *controlling the X and Y motion of a milling machine using sets of coordinates representing machining path data* in 1948. In June of 1949, the Air Force awarded a contract to Mr. Parsons to convert a tracer type milling machine to an NC machine. Parsons in turn subcontracted the project to MIT servomechanisms laboratory to design servomechanism (self-regulating feedback system) for machine tools. In 1952, NC came into existence at MIT as a retrofitted vertical spindle Cincinnati Hydrotel Tracer Milling Machine. It used a binary perforated tape as the medium for storing the machining program. Mr. Parson was fired by his own company for his dedication and persistence in following through the radically new concept. However, President Regan honored him in 1988 with a National Medal of Technology. Further development was due to advances in digital electronics and microprocessors. The first commercial NC machines were exhibited at the 1955 National Machine Tool Show. The first generation of NC machines used large vacuum-tube based controllers. Second generation models replaced the vacuum tubes with transistors. Third generation models featuring integrated circuitry and modular circuit design reduced costs and increased reliability further [1].

A. Sirinterlikci and Y. Ertekin, *A Comprehensive Approach to Digital Manufacturing*, Synthesis Lectures on Mechanical Engineering, https://doi.org/10.1007/978-3-031-25354-6_3

3.2   Operating Principle of NC

Numerical Control/Computer Numerical Control (NC/CNC) is a specialized and versatile form of *soft (programmable) automation*, and its applications cover many kinds, although it was initially developed to control the motion and operation of machine tools. CNC may be a means of operating a machine using discrete numerical values fed into the machine, where the required 'input' technical information is stored on a kind of input media such as floppy disk, hard disk, CD ROM, DVD, USB flash drive, or RAM card etc. The machine follows a predetermined sequence of machining operations at the predetermined speeds necessary to produce a workpiece of the right shape and size, and thus completely yields predictable results. A different product can be produced through reprogramming and a low-quantity production run of different products is justified. There are many industrial operations in which the position of a work head must be controlled relative to the part or product being processed.

Two categories of NC applications:

1. Machine tool applications
2. Non-machine tool applications

CNC technology is widely used for machining operations such as turning, drilling, and milling. The technology has motivated the development of machining centers, which change their own cutting tools to perform a variety of machining operations under NC. Other types of CNC machine tools include: Grinding machines, sheet metal press-working machines, tube bending machines, EDM (Wire or Sinker) (Fig. 3.1).

Non-machine tool applications encompass equipment like tape laying and filament winding machines for composites (Fig. 3.2), welding machines (arc and resistance welding), and plasma cutting. Component insertion machines in electronics assembly, drafting machines, coordinate measuring machines (CMMs) for inspection, fabric and laser cutting machines (Fig. 3.2), and 3D Printers are also CNCs.

Fig. 3.1 Wire EDM (top left), CNC-Lathe (top middle) CNC, Umbrella type automatic-tool-changer (ATC) for a vertical machining center (VMC) (top right), VMC and turning center (TC) (bottom left and middle), Emco Concept Turn 250 TC (Drexel University Machine Shop), 5-Axis CNC Router (bottom right)

Fig. 3.2 Tape laying machines for aerospace applications (left and middle), ultrasonic laminate cutting (right) (Courtesy—Cincinnati Machine Tools)

3.2.1 Control Systems

Control systems of the CNCs are of two different types:

- Open loop systems
- Closed loop systems

Open loop (Fig. 3.3) systems have no access to the real-time data about the performance of the system and thus no immediate corrective action can be taken in case of system disturbance [1]. This system is normally applied only to the case where the output is almost constant and predictable. Therefore, an open loop system is unlikely to be used to control machine tools since the cutting force and loading of a machine tool is never constant.

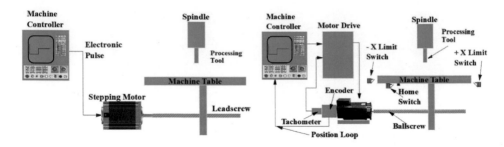

Fig. 3.3 Block diagram of an open (left) and closed loop (right) system

The only exception is the wire cut machine for which some machine tool builders still prefer to use an open loop system because there is virtually no cutting force in wire cut machining.

In a closed loop system (Fig. 3.3), feedback devices closely monitor the output and any disturbance will be corrected in the first instance [1]. Therefore, high system accuracy is achievable. This system is more powerful than the open loop system and can be applied to the case where the output is subjected to frequent change. Because DC servomotors can be instantly reversed and have higher torque, permitting to handle heavier loads than stepping motors, they are used in closed-loop systems for the larger NC machines as well as for many of the smaller NC machines. Nowadays, almost all CNC machines use this type of control system.

3.3   CNC Turning and Machining Centers

Modern turning centers have advanced considerably in capability, flexibility, versatility, and reliability. Because of their increased flexibility and capability, CNC lathes are classified in two types: vertical and horizontal. Vertical NC turning centers are modern adaptations of the manual vertical turret lathes (VTLs), also including tool change and part loading/unloading capabilities. Horizontal CNC turning centers of the shaft, chucker, or universal type have not only changed relative to advanced features and technology, but in basic construction as well. Many modern horizontal CNC turning centers are of the slant-bed design (Fig. 3.1, Emco Concept Turn 250) include easy access for loading, unloading, and measuring, allowance for chips to fall free, minimum floor space utilization, ease and quickness of tool-changes, and better strength and rigidity. Four-axis and dual-spindle turning centers have also gained considerable acceptance among manufacturers. Four-axis lathes (Fig. 3.4, Okuma TC) provide sizable savings and productivity increases over conventional two-axis machines because Outside Diameter (OD) and Inside Diameter (ID) operations can be performed simultaneously through independently programmed slides. Dual-spindle machines have two spindles with two independent slide

Fig. 3.4 Top row: FEA analysis on CNC machine structure, Cincinnati machine H5 1000AP 5-Axis High Speed Horizontal Machining Center, Cincinnati Machine HyperMach Rail-Type Gantry Profiler for aerospace applications. Bottom row: FMS for engine block/heads machining, Okuma Mill-Turn 4-Axis Turning Center (Multus B400) Finished Fuselage Bulkhead machined on Hyper-Mach

motions for OD and ID operations for each spindle. Machines of this type are also capable of achieving high productivity levels with considerable savings. Operations performed on CNC turning centers are basically no different than those performed on older or conventional machines. These consist of the standard turning, facing, drilling, boring, tapping, and threading. These modern CNC versions can still remove metal no faster than their conventional counterparts; however, enhanced cutting tool technology, tool-changing and work-holding methodology, and new automated features have considerably shortened the non-cutting time.

Machining centers existing in the 1960s had automatic tool changers (ATCs), allowing machines to perform a variety of machining operations on a workpiece by changing their own cutting tools. Thus began a tool change and additional feature/capability revolution among machine tool builders that continues to escalate by adding improvements and enhancements to the staggering array of machining center choices.

Vertical Machining Centers (VMC) continue to be widely accepted and used, primarily for flat parts and where three-axis machining is required on a single part face such as in mold and die work. Horizontal Machining Centers (HMC) are also widely accepted and used, particularly with large boxy, and heavy parts and because they lend themselves to easy and accessible pallet shuttle transfer when used in a cell or Flexible Manufacturing System (FMS) application (Fig. 3.4).

Machining center construction (Fig. 3.5) has improved to accommodate higher spindle speeds, feeds, and horsepower requirements, along with overall higher utilization rates and increased performance requirements:

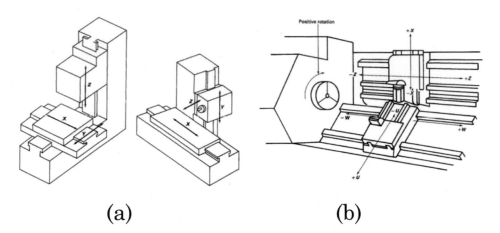

Fig. 3.5 Axes designation for VMC and HMC (**a**), TC (**b**) [2]

- Five axis (and more) of movement: key to new designs are pitch and roll motions right in the spindle head
- Horizontal and vertical spindle: these are similar in appearance to right-angle spindle
- Attachments that have long been available to change the spindle orientation by 90°.
- An ATC: tool-storage mechanisms (Fig. 3.1) vary among the diversified machine tool suppliers, as some are front, side, or top mounted.
- High-technology carbide-insert cutting tools are gaining wide acceptance and prominence due to their inherent accuracy and ease of maintainability. This eliminates removing the entire tool from the holder and re-sharpening.

3.3.1 Axes Designations

The primary translational axes of both vertical and horizontal machining centers are *X, Y and Z (Additional axes represent rotation about these translational axes) (Fig. 3.5)*. On VMCs, the X axis provides the longitudinal table travel, the Y axis provides in and out saddle movement, and the Z axis provides up and down movement of the head or spindle [2]. On HMCs X-axis movement is also through the longitudinal table travel. Y-axis movement is up and down, provided through movement of the machine tool's knee or spindle carrier. Z-axis positioning is through in and out movement of the machine tool's saddle, table, or spindle carrier [2]. HMCs also provide Beta-axis movement, which greatly increases versatility. Beta-axis can be used to index a workpiece for machining in the X, Y and Z plane or simultaneously with one or more other axes for a contouring cut.

The primary axes of a turning center are Z and X. The Z axis travels parallel to the machine spindle, while the X axis travels perpendicular to the machine spindle. *U and W* axes are typically auxiliary axes providing additional movement (*in an incremental fashion*) and tool capacity. For both OD and ID operations, a negative Z is a movement of the saddle toward the headstock. A positive Z is a movement of the saddle away from the headstock. A negative X moves the cross slide toward the spindle center-line, and a positive X moves the cross slide away from the spindle center-line.

3.4 Introduction to CNC Programming

Basic Geometrical Factors

Workpiece Coordinate System
So that the machine can work with the specified positions, the data must be given within a reference system corresponding to the directions of movement of the axis carriages. For this purpose, a coordinate system with X, Y and Z axes is used. As specified in DIN 66217, *right-handed and orthogonal (cartesian) coordinate systems* are used for machine tools.

Workpiece Zero Point
The workpiece is now represented in this coordinate system. In this system, the user needs to work with both positive and negative position data. Usually, all data within the coordinate system are related to the origin (X = 0, Y = 0, Z = 0). However, in NC programming, the origin is always called the workpiece (part or program) zero (W) (Fig. 3.6). It is also called by OEMs, work-offset.

Determining Workpiece Positions

Description of Positions on the X, Y Axes
Now the user has to apply an imaginary scale to the coordinate axes so that one can uniquely identify each point in the coordinate system by a direction (X, Y, Z) and three numerical values. In the following examples, the workpiece zero always has the coordinates X = 0, Y = 0, Z = 0.

In-Practice
Example review:
   For simplification, this example considers only one plane of the coordinate system, i.e. the X/Y plane. Points P1 to P4 therefore have the following coordinates (Fig. 3.7, Left):

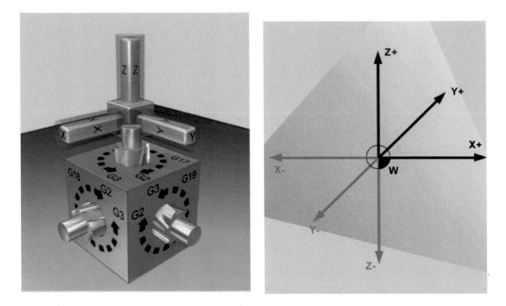

Fig. 3.6 The illustration shows the positions where workpiece zero is usually located

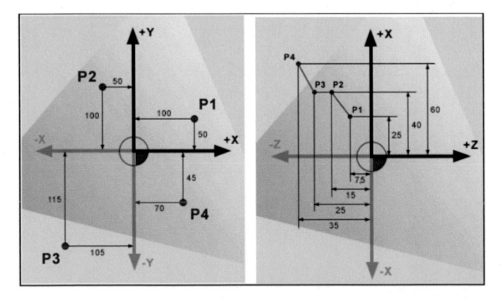

Fig. 3.7 Determining DIN ISO interpolation planes and position coordinates. (Left: mill example, Right: turning example)

P1 corresponds to X100 Y50, P2 corresponds to X-50 Y100, P3 corresponds to X-105 Y-115, P4 corresponds to X70 Y-45.

For lathes, one plane is usually sufficient to describe the profile (Fig. 3.7, Right):

Example review: Points P1 to P4 are determined by the following coordinates:

P1 corresponds to X25 Z-7.5, P2 corresponds to X40 Z-15, P3 corresponds to X40 Z-25

P4 corresponds to X60 Z-35

Definition of Depth Setting

For milling machines, the depth setting must be defined. User therefore has to assign a numerical value to the third coordinate. (In this case Z).

In Practice

Example review: Points P1 to P4 are determined by the following coordinates: P1 corresponds to X10 Y45 Z-5, P2 corresponds to X30 Y60 Z-20, P3 corresponds to X45 Y20 Z-15 (Fig. 3.8, left)

Description of Position by Polar Coordinates

The method described above for determining points within the coordinate system is called the "Cartesian coordinates". There is also another way of specifying the coordinates,

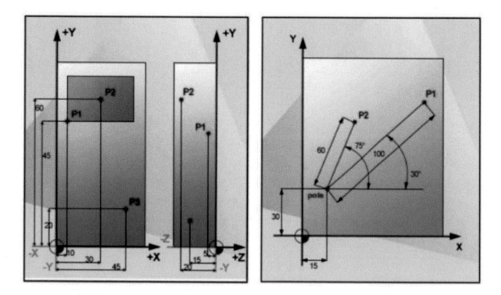

Fig. 3.8 Definition of depth setting for milling (left), positions by polar coordinates (right)

which is called the "polar coordinates". Polar coordinates are useful when a workpiece
is dimensioned by a radius and an angle. The point from which dimensioning starts is
called the "pole" (the angle is positive (+) if counterclockwise, and negative if clockwise
(−)). Figure 3.8 shows an example of this type of dimensioning. In this example, points
P1, P2 have the following positions: P1 corresponds to radius = 100 and angle = 30°,
P2 corresponds to radius = 60 and angle = 75° (Fig. 3.8, right).

Absolute Dimensioning
In the previously described method for determining positions, all position data are related
to the current work zero point. For tool movements, this means that: the absolute
dimension value indicates where the tool is to move to starting from zero point.

In Practice
Example (Fig. 3.9 [3]) **review:**
    The position data are as follows:
    P1 corresponds to X20 Y35, P2 corresponds to X50 Y60, P3 corresponds to X70
Y20.

Chain (Incremental) Dimensioning

Beside the absolute dimension, "chained dimensions" are another basic way of specifying positions. In this case, all position data are related to the last point specified.
This type of dimensional notation is also called "incremental dimension".
    The incremental dimension value indicates the distance the tool is to move start-
ing from the last point (Fig. 3.9b) [3]. In this example, P4 corresponds to X10 Y10;

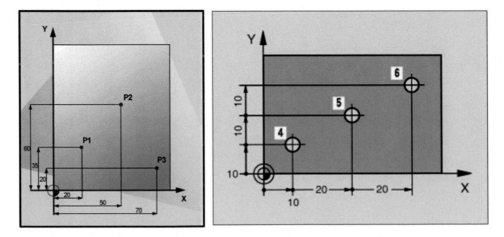

Fig. 3.9 Absolute dimensions example (**a**), Chain dimensions example (**b**) [3]

relative to zero, P5 corresponds to X20 Y10; relative to P4, and P6 corresponds to X20 Y10; relative to P5.

Predefined Points in the Working Area: Reference Points
Certain points in the work area are already predefined on the NC machine. These can be subdivided into (Fig. 3.10): Point at which the machine is to start, i.e. R: reference point, Point to which programming of the workpiece dimensioning relates: W: workpiece zero (part or program zero) point, B: starting point, A: stop point. Point defined by the machine manufacturer: M: machine zero point (Machine Home position).

Machine Datum Point M
The machine datum (a.k.a. Machine Zero/Home) point is the zero-point defined by the machine tool manufacturer. The machine datum point is the reference point for measurements for the whole machine.

For lathes, the datum point lies at the point of intersection between the axis of rotation and the plane surface of the main spindle; for milling machines, it is the clamping surface of the workpiece (Fig. 3.11).

Dead Stop Point A
The dead stop point is the point where the workpiece is attached. For turning lathe, this is usually the face surface of the lathe chuck; for milling machines, it is the fixed jaw of the vise (Fig. 3.12).

Fig. 3.10 Predefined points in the work area (lathe example) [3]

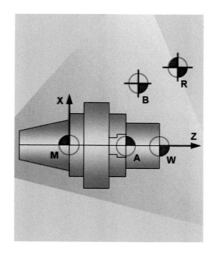

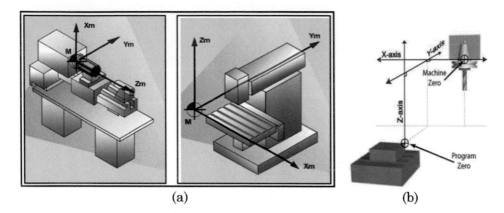

Fig. 3.11 Machine datum point (M) for CNC lathe (**a**) and CNC mill (**b**) [3]

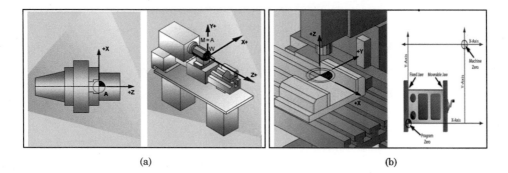

Fig. 3.12 Dead stop point A for CNC lathe (**a**) and CNC mill (**b**) [3]

Workpiece Zero-Point W

Workpiece (part or program) zero is the origin of the workpiece coordinate system and can be selected as desired by the programmer. Zero-point shifts or frames can be used to move the workpiece zero position within the NC program (Fig. 3.13).

Retaining (Gage) point F and Offsets (Fig. 3.14)

Retaining (Gage) point F is located at the tool holder and is the reference point from which tool point P is measured.

Tool Point P (Control Point)

Tool point P (or B) is the offset point calculated to the tool cutting edge. For measuring the tool, the distances from the tool point to the toolholder retaining point are determined. These values are then input to the tool adjustment memory in the CNC control system and

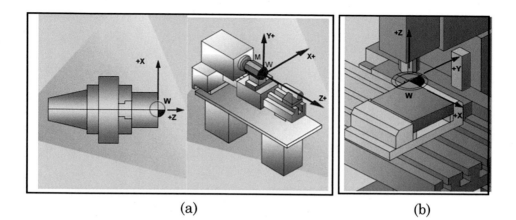

Fig. 3.13 Workpiece zero point (W) for CNC lathe (**a**) and CNC mill (**b**)

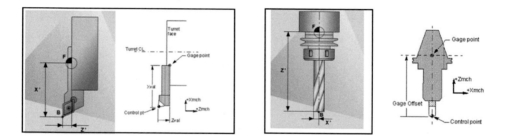

Fig. 3.14 Retaining (Gage) point F and offsets

can be retrieved at address i.e. T04 D04 H04 (where *D and H registers represent diameter and height (tool length) offset of tool #4*).

Cutting Points

The CNC controller needs to know the position of cutting point P relative to radius center point S so that it can calculate the correct tool path for creating the programmed contour. The cutting-edge position S defines the position of the center of the cutting-edge radius S relative to the starting point B. The CNC controller can only perform the cutting radius compensation once the cutting radius and cutting position have been specified.

For turning lathe tools there are nine possible cutting positions (Fig. 3.15, right), but for milling tools (Fig. 3.15, left), since the tool retaining point and cutting radius center point are the same, there is only one cutting position which does not have to be defined.

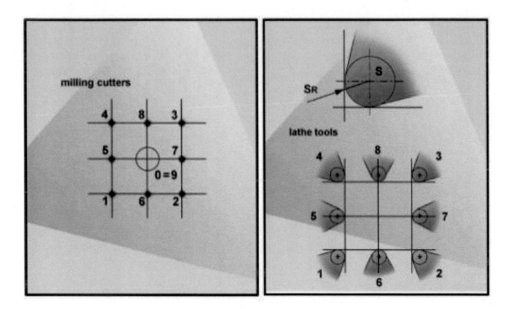

Fig. 3.15 Cutting points [3]

Position of the Coordinate Systems

For programming machine tools, it is useful to apply various coordinate systems. Machine zero and program (part) zero each have their own distinct coordinate systems. Machine zero is associated with the machine coordinate system, while program zero is associated with the work coordinate system (Fig. 3.11b). These are usually distinguished by their origin points: The origins of both coordinate systems ($MX_mY_mZ_m$ and $WX_wY_wZ_w$) are shown together in Fig. 3.16.

- Machine coordinate system (origin = M)

  After approaching the reference point the NC position displays of the axis coordinates are relative to the machine zero point (M) of the machine coordinate system (MCS).

- Workpiece coordinate system (origin = W)

  The program for executing the workpiece is relative to the workpiece (program or part) zero point (W) of the workpiece coordinate system (WCS). Machine zero point and workpiece zero point are usually not identical. The distance between the points is the entire zero-point offset and is made up of various shifts.

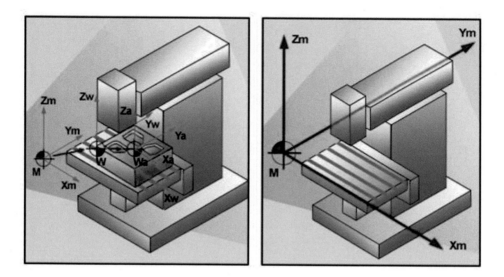

Fig. 3.16 Position of the coordinate systems (milling examples)

Like machine zero, the machine coordinate system is absolute and can never be changed. The machine coordinate system is the benchmark location to which all else is referenced. Its X-axis and Y-axis origin is located at the upper right corner of the mill table. The Z-axis origin is above the work area, a few inches below automatic tool change position. Because of this orientation, all machining performed on a part is done in negative machine coordinates, relative to machine zero. The machine coordinate system is made up of all physically present machine axes. Among others, the reference points and tool change positions are defined in the machine coordinate system. If the programming is performed directly in the machine coordinate system, the physical axes of the machine are addressed directly.

Like program or part zero, the work coordinate system defines the location of the workpiece and may be changed by the programmer and operator. The work coordinate system has program zero (W) as its origin.

The Three-Finger Rule (Right Hand Rule)
The position of the cartesian coordinate system relative to the machine depends on the type of machine. The axial directions follow the "three finger rule" (as defined in DIN 66217). If you stand in front of the machine with the middle finger of your right hand pointing against the feed direction of the main spindle, then the thumb indicates direction X+, the index finger indicates direction Y+, the middle finger indicates direction Z+.

If you apply the three-finger rule to various machine types, you will get different results. The illustration in Fig. 3.17 shows some examples of machine coordinate systems [3].

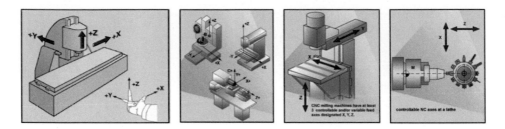

Fig. 3.17 The three-finger rule and axis designation examples

Workpiece Coordinate System
The workpiece coordinate system describes the workpiece geometry. In other words, the data in the NC program refer to the workpiece coordinate system. The workpiece coordinate system is typically a cartesian coordinate system assigned to a certain workpiece.

Considerations for Selecting Program Zero
Theoretically, program (part) zero can be located anywhere within the machine coordinate system. However, some areas of the part are better than others. As a programmer, one should recognize effective locations for program zero. Machining on a CNC mill is performed on a workpiece clamped on a machine table with a vise or fixture. Ideally, the program zero edge should be against a fixed location, such as the fixed jaw of the vise, as shown in Fig. 3.18. Inexperienced programmers may incorrectly use a moving jaw as the reference edge. Having program zero on a moveable jaw can produce inaccurate machining results unless your blank material is 100% identical for all parts.

For the z-axis, a common practice is to select the top face of the finished part. This makes the Z-axis positive above the face and negative below the face. It is best to have all dimensions described in the part program in the first quadrant, as shown in Fig. 3.18.

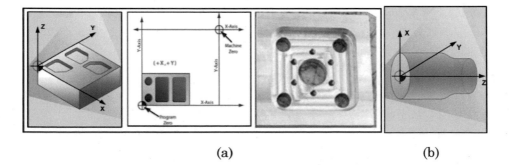

(a)   (b)

Fig. 3.18 Workpiece coordinate system **a** mill, **b** lathe

Mistakes are more likely made when dealing with negative part program values. Notice also that the part reference edge is against the fixed Jaw. For round parts or patterns (bolt circles, circular pockets), program zero is best located at the center of the part, as shown In Fig. 3.18. By locating program zero in the center of circular and symmetrical parts, the programmer has a mirror image of each side of the part, with opposite matching points that are the same distance from the center.

Definition of Programming
NC programming is where all the machining data is compiled and where the data is translated into a language which can be understood by the control system of the machine tool. The machining data is as follows:

- Machining sequence, tool start up point, cutting depth, tool path (based on the geometry selected) etc.
- Cutting conditions spindle speed, feed rate, coolant, etc.
- Selection of cutting tools.

Word Address Format
The word address format features an address for each data element to permit the controller to assign data to their correct register in whatever order they are received. A single alpha character (said to be a "word") is used to identify each register. The Electronics Industries Association (EIA) has developed and the American National Standards Institute (ANSI) has adopted a standard format of NC data, ANSI/EIA RS-274-D [4]. Most (but not all) NC controllers use the address words F, G, I, J, K, M, N, S, T, X, Y, and Z. The more complex (i.e. 5 axis) machines have more functions to perform, therefore more registers are required and additional address words such as A, B, and C are required to indicate rotation about the X, Y, and Z axes. Word address formatting permits unneeded commands to be omitted; it also permits several miscellaneous and preparatory function commands to be put in a single block rather than in several successive blocks. The ANSI/EIA standard RS-274-D [4] establishes a method to specify word address data. This data format specification is used to indicate:

1. Which address words are used by a given controller,
2. The number of digits required for each register,
3. Where the decimal point is placed.

The ANSI/EIA standard RS-274-D word format: [4]
    N3.0 G2.0 XYZIJK3.4 F3.2 S4.0 M2.0
    This would mean that the sequence numbers (N) have 3 digits to the left of the decimal and zero digits to the right. The X, Y, Z, I, J, and K commands all have 3 digits

to the left of the decimal and 4 digits to the right, etc. The word address format, in theory, should permit the programmer to enter command data into a block in any order desired, but it usually doesn't work out that way. Most controllers require commands to be entered in a specific order, usually in the RS-274-D order [4]. Some controllers vary the order somewhat and the number of digits before and after the decimal can vary from controller to controller. The programmer must always consult the programming manual for the particular NC machine to determine the required order and word format required by its controller.

The most common 'addresses' are listed below: [4]

Function	Address
• Sequence number	N
• Preparatory function	G
• Coordinate word	X, Y, Z
• Parameters for circular interpolation	I, J, K
• Feed function	F
• Spindle function	S
• Tool function	T
• Miscellaneous function	M

An example of a program segment is as follows:
N20 G01 X20.5 F200 S1000 M03
N21 G02 X30.0 Y40.0 I20.5 J32.0

Some G words alter the state of the machine so that it changes from cutting straight lines to cutting arcs (G02 or G03). Other G words cause the interpretation of numbers as millimeters (G21 or G71) rather than inches (G20 or G70). While still others set or remove tool length or diameter offsets (G49 and G40). Most of the G words tend to be related to motion or sets of motions. Appendix I lists the currently available G and M words from EIA Standard RS274D [4].

Guideline for the Structure of a NC Part Program
An (NC/parts) program consists of a sequence of NC blocks (see Fig. 3.19). Each block represents one processing step. In one block, instructions are written in the form of words. Each program is closed by a special word for end of program, i.e. M2, M17 and/or M30.

Language Elements—Milling
Just as in spoken language, an NC program consists of sentences, which in this case are called blocks. These blocks in turn are made up of words. A word in "NC language" consists of an address character (usually an alphabetical letter) and a digit and/or a string of digits representing an arithmetic value. The string of digits may contain signs and decimal points. The sign is always placed between the address characters and the string of

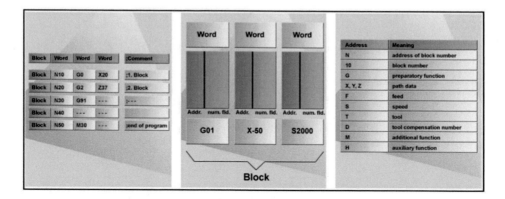

Fig. 3.19 NC part program structure using word address format [4]

digits. Positive signs (+) do not need to be written. All words in a block are performed simultaneously. Blocks are separated by the *End-Of-Block (EOB)* character—usually generated by pressing the return key. The EOB character tells the controller to execute the commands contained in the block and signals the start of the next block. Blocks are made up of one or more commands.

Sequence of Words in a Single Block
To create a clear block structure, the words in a block should be arranged schematically (see Fig. 3.19). Common syntax:

- First, all G-functions
- Then the coordinates X, Y or Z
- Then any other commands, such as S, T, F…
- At the end of the block there are the M functions.

Reference Block/Subblock
Some addresses may be used several times within one block (e.g. G…, M…, H…).
    There are two types of blocks:

Reference Blocks
A reference block must contain all words that are required for starting the working sequence from the program section that starts with the reference block. Reference blocks may occur both in main programs and in subroutines. The control system does not check if a reference block contains all the information required. Identifying a block as a reference block is important for reference block searches or searches starting from the last reference block.

Subblocks
A subblock contains all the information required for the relevant operating step.

Block Number
Reference blocks are marked by a reference block number. A reference block number consists of the character ":" and a positive integer (block number). The block number is always placed at the beginning of the block. Reference block numbers must be clearly identifiable within a program, in order to obtain a definite result when a search is made.

> **In Practice**
> **Example review:**
>   :10 T2 G96 S180 F0.2 M4.
>   Subblocks are marked by a subblock number. A subblock number consists of the character "N" and a positive integer (block number). The block number is always placed at the beginning of the block.
>   **Example review:** N50 G00 X1 10 Y0 Z3 M08
>   Subblock numbers must be clearly identifiable within a program, in order to obtain definite search results. The order sequence of block numbers is optional. However, ascending order block numbers are recommended. NC blocks can also be programmed without block numbers.

Travel Commands (G-commands)
Basically, we can distinguish between the following travel commands:

1. G0 (or G00) Positioning of the tool with rapid feed (highest possible feed rate) on a straight line
2. G1(or G01) Positioning of the tool with defined feed rate on a straight line (linear interpolation)
3. G2 (or G02) Positioning of the tool with defined feed rate in a circle in clockwise direction (circular interpolation)
4. G3 (or G03) Positioning of the tool with defined feed rate in a circle in counterclockwise direction (circular interpolation)

Using a G0 in your code is equivalent to saying, "go rapidly to point xxx yyyy". This code causes motion to occur at the maximum available traverse rate (Fig. 3.20).

> **In Practice**
> **Example review:**
>   N100 G0 X10.00 Y5.00

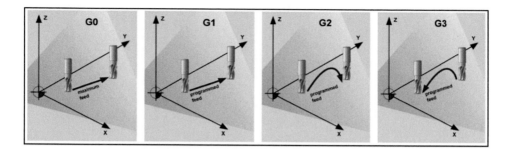

Fig. 3.20 Travel commands (preparatory function G-commands) examples

This line of code causes the spindle to rapid travel from wherever it is currently to coordinates X = 10″, Y = 5″

G1 causes the machine to travel in a straight line with the benefit of a programmed feed rate (using "F" and the desired feed rate). This is used for actual machining and contouring.

Example review:

N120 Z0.1 F6.0—move the tool down to Z = 0.1 at a rate of 6 in./min

N130 Z-0.125 F3.0—move tool into the workpiece at 3 in./min

N140 X2.5 F8.0—move the table, so that the spindle travels to X = 2.5 at a rate of 8 in./min. Note that all travel commands are executed with respect to tool motion.

G2 causes clockwise circular motion to be generated at a specified feed rate (F). The generated motion can be 2-dimensional, or 3-dimensional (helical). On a common 3-axis mill, one would normally encounter lots of arcs generated for the X, Y plane, with Z axis motion happening independently (2 axis moves in G17 plane). But, the machine is capable of making helical motion, just by mixing Z axis moves in with the circular interpolation. When coding circular moves, you must specify where the machine must go and where the center of the arc is in either of two ways: By specifying the center of the arc with I and J words (Fig. 3.21), or giving the radius as an R word. I is the incremental distance from the X starting point to the X coordinate of the center of the arc. J is the incremental distance from the Y starting point to the Y coordinate of the center of the arc.

In Practice

Example review:

G1 X0.0 Y1.0 F20.0—go to X0.0, Y1.0 at a feed rate of 20 in./min

G2 X1.0 Y0.0 I0.0 J-1.0—go in an arc from X0.0, Y1.0 to X1.0 Y0.0, with the center of the arc at X0.0, Y-1.0 (Incremental I0.0 and J-1.0 distances from arc start point). G3 is the counterclockwise sibling to G2.

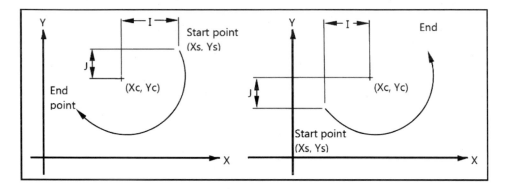

Fig. 3.21 Circular interpolation—clockwise (G2) commands (left) and Counterclockwise (G3) (right)

Fixed and Variable Identifiers

Addresses are fixed or variable identifiers, e.g. for geometric axes (X, Y,..) spindle axes (S) or rotary axes (A, B, C) and/or any other system parameters (interpolation parameters I, J, K) (Table 3.1).

Modal/Block-Serial Addresses

Many G codes and M codes cause the machine to change from one mode to another, and the mode stays active until some other command changes it implicitly or explicitly. Such commands are called "*modal*". Modal addresses remain valid keeping their programmed value (in all subsequent blocks) until a new value is programmed under the same address.

Example

G01 X10 Y10 F0.2 Z0

Block-serial addresses apply only to the block in which they are programmed.

Example

G53 G0 T0 D0 X200 Z200

Modal codes are like a light switch. Flip it on and the lamp stays lit until someone turns it off. For example, the coolant commands are modal. If coolant is turned on, it stays on until it is explicitly turned off. The G codes for motion are also modal. If a G01 (straight move) command is given on one line, it will be executed again on the next line unless a command is given specifying a different motion (or some other command which implicitly cancels G01 is given). "Non-modal" codes affect only the lines on which they occur. For example, G04 (dwell) is non-modal.

Modal commands are arranged in sets called "modal groups". Only one member of a modal group may be in force at any given time. In general, a modal group contains

Table 3.1 Fixed and variable identifiers [4]

Address	Meaning
A	Rotary axis
B	Rotary axis
C	Rotary axis
D	Cutting edge number
F	feed
H	Auxiliary function
I	Interpolation parameter
J	Interpolation parameter
K	Interpolation parameter
L	Invoke of subroutine
M	Additional function
N	Subblock
P	Number of program runs
R	operands
S	spindle speed
T	Tool number
X	Axis
Y	Axis
z	Axis
:	Reference block
Fixed addresses	
Address	Meaning
D	Cutting edge numbers
F	feed
H	Auxiliary function
L	Invoke of subroutine
M	Additional (Miscellaneous) function
N	Subblock
P	Number of program runs
R	Operands
S	Spindle speed
T	Total number
:	Reference block

Table 3.2 Modal groups

Group 1	{G0, G1, G2, G3, G80, G81, G82, G83, G84, G85, G86, G87, G88, G89}—motion
Group 2	{G17, G18, G19}—plane selection
Group 3	{G90, G91}—Absolute, Incremental distance mode
Group 5	{G93, G94}—spindle speed mode
Group 6	{G20, G21}—units
Group 7	{G40, G41, G42}—cutter diameter (radius) compensation
Group 8	{G43, G49}—tool length offset
Group 10	{G98, G99}—return mode in canned cycles
Group 12	{G54, G55, G56, G57, G58, G59, G59.1, G59.2, G59.3} coordinate system selection
Group 2	{M26, M27}—axis clamping
Group 4	{M0, M1, M2, M30, M60}—stopping
Group 6	{M6}—tool change
Group 7	{M3, M4, M5}—spindle rotation
Group 8	{M7, M8, M9}—coolant
Group 9	{M48, M49}—feed and speed override bypass

commands for which it is logically impossible for two members to be in effect at the same time. Measurement in inches vs. measure in millimeters are modal. A machine tool may be in many modes at the same time, with one mode from each group being in effect. The modal groups are shown in Table 3.2.

Types of Path Data
In the CNC program, path data commands can be inserted as desired. Basically, we can distinguish between the following types of path data:

1. Absolute programming (G90), absolute dimension notation
2. Chained (incremental) dimension programming (G91), relative dimension notation
3. Metric/inch dimension data (G20, G21 for Fanuc Controls. Siemens controls use G70 and G71). Depending on the dimension entries in the production drawing, you can program workpiece-related geometric data either in metric or inch dimensions.

   Example G code segment for Fig. 3.22a: N10 G21 G90 G41 X10. Y50. F254.
   Example G code segment for Fig. 3.22a: N10 G21 G90 G41 X60. Y85.
   Example G code segment for Fig. 3.22b: N10 G21 G90 G41 G01 X10. Y50. F254.
   N20 G91 X50. Y35.
   Example G code segment for Fig. 3.22c: N10 G20 G90 G40 G01 X1.18 Y50. F10.
   N20 G21 X90. Y30.

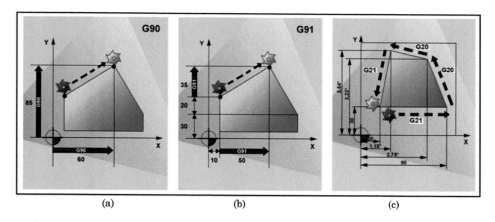

Fig. 3.22 Absolute (G90) and incremental (G91) programming

N30 G20 X3.22.
N40 X1.18 Y3.54.
N50 G21 X20. Y30.

Absolute Programming (G90)
Command G90 is used to define the description system for moving to the target positions.
The dimension data refer to the origin point of the currently applied coordinate system.
You program the point to which the tool is to move, e.g. in the workpiece coordinate
system. Command G90 is generally effective for all axes programmed in the following
NC blocks. It is a modal command.

In Practice

Example program review:

The travel distances are entered as absolute coordinates relative to the workpiece
zero point. Center coordinates I and J for the circular interpolation are programmed
as chained dimensions—independent of G90 (Fig. 3.23).

N10 T1 S2000 M3;	Tool, rotational speed (rpm), spindle one right
N20 G90 G0 X45 Y60;	Absolute data entry, rapid feed to position XYZ
N30 G1 Z-5 F500;	Set the tool advance
N40 G2 X20 Y35 I0 J-25;	End of circle as absolute dimension X, Y, circle center as an increment from starting point I (for X) J (for Y)
N50 G0 Z2;	Move out
N60 M30;	End of program

Fig. 3.23 Absolute (G90)
programming with circular
interpolation

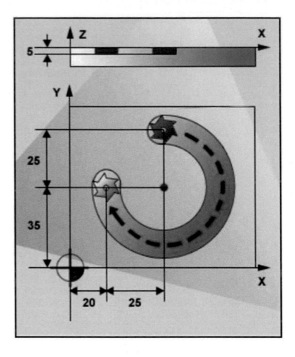

Incremental Programming (G91)

Command G91 is used to define the description system for moving to the target positions. The positional data refers to the last point occupied. You program the distance the tool is to move from its current location. In general, command G91 is effective for all axes programmed in the subsequent NC blocks. The command is modal.

The travel distances are entered as absolute coordinates relative to workpiece (program) zero point. Center coordinates I and J (Fig. 3.21) for circular interpolation are specified as incremental coordinates starting from the circle starting point, as the circle center is usually programmed as an incremental dimension—independent of G91:

N10 T1 S2000 M3;	Tool, rpm, spindle one right
N20 G91 G0 X45 Y60 Z2;	Absolute data entry, rapid feed to position XYZ
N30 G1 Z-7 F500;	Tool set in forward feed
N40 G2 X-25 Y-25 I0 J-25;	End of circle as incremental dimension
N50 G0 Z7;	Move out
N60 M30;	End of block

Selection of Working Planes (G17-G19)
In the basic setting (default state), the correct plane is pre-defined for each machine. It must be programmed only if the plane must be re-positioned.

- G17: plane X/Y
- G18: plane X/Z
- G19: plane Y/Z

By specifying the working plane in which the desired contour is to be produced, the following functions are defined simultaneously:

1. Plane for tool radius compensation
2. Tool advance direction for tool length compensation depending on the type of tool
3. Plane for circular interpolation

It is recommended to define the working plane at the start of the program. When calling tool radius compensation G41/G42, the working plane must be specified so that the control system can adjust the tool length and radius. In the basic setting, G17 (X/Y plane) is always the default. The 'conventional' procedure is as follows (Figs. 3.24 and 3.25):

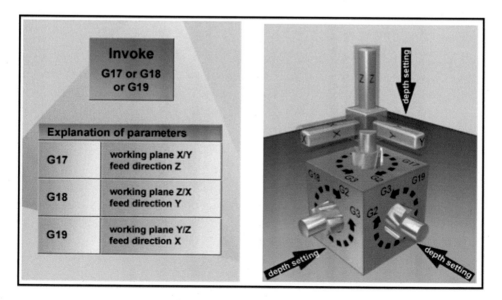

Fig. 3.24   Selection of working planes (G17-G19)

List of predefined M-functions	
M0*	programmed stop
M1*	optional stop
M2*	end of main program and returning to start of program
M30*	end of program, like M2
M17*	end of subroutine
M3	spindle right-handed rotation
M4	spindle left-handed rotation
M5	spindle stop
M6	tool change (standard setting)

Fig. 3.25 Commonly used M codes

In Practice

Define working planes, call tool type and tool compensation values, start path compensation. **Example:**

Program travel movements:	Comments:
N10 G17 T5 D8;	G17 Call working planes, in this case X/Y. T, D call tool Length compensation is performed in Z-direction.
N20 G1 G41 X10. Y30. Z-5. F500.;	Radius compensation is performed on the X/Y plane.
N30 G2 X22.5 Y40. I50. J40.;	Circular interpolation with tool radius compensation on X/Y plane.

Machine Switching Commands (M-Commands)

The M (miscellaneous) functions (M… possible values 0–999 999, integer) can be used to activate switching operations such as "Coolant ON/OFF" and other machine functions. A small number of M functions are already assigned to fixed functions by the manufacturer of the control system.

All unassigned M function numbers may be assigned by the machine manufacturer, e.g. to a switching function for controlling the clamping devices or for switching other machine functions on/off.

The following applies to NC blocks in which a travel command and an M-command are programmed. The machine data define whether the programmed M-command of the current NC block is executed: before the travel movement, along with the travel movement, after the travel movement. The commands M0, M1, M2, M17 and M30 are always executed after the travel movement. Some M-functions which are important for program flow are already pre-assigned in the standard version of the control system:

1. Programmed stop, M0

   In the NC block with M0, processing is stopped. Chips can then be removed, measurements checked, etc.

2. Optional stop, M1

   Similar to M0, M1 needs additional machine parameters to be set for the program to stop.

3. End of program, M2, M17, M30

   A program is ended by M2, M17 or M30 and set back to program start. If the main program is called (as a subroutine) by another program, M2/M30 acts as an M17 and vice versa, i.e. M17 has the same effect in the main program as an M2/M30.

4. Spindle functions M3 (clockwise/forward rotation), M4 (counter- clockwise/reverse rotation), M5 (spindle stop)

Example program

| N10 G0 X50. S1000 M3; | M-functions in the block with axis movement, spindle starts up before X-axis movement |
| N180 G1 X10 M4 M8; | 5 M-functions max. in one block |

Metric/Inch Dimension Data (G20, G21)

Depending on the dimensions entered in the production drawing, user can program workpiece-related geometric data either in metric or inch dimensions. The following geometric data can be converted by the control system (with the necessary deviations) into the measurement system not currently selected, and can therefore be entered directly:

- path information X, Y, Z, …
- thread pitch
- programmable zero-point shifts

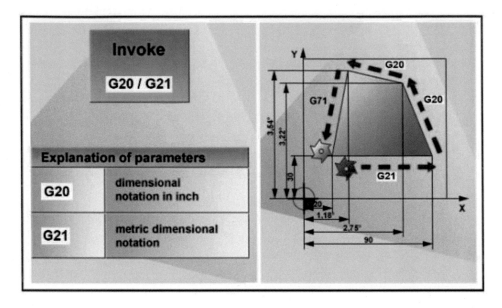

Fig. 3.26 Inch versus Metric dimension data (FANUC controls)

In Practice

Example program (Fig. 3.26).

Changing between inch dimension entries and metric dimension data with the metric basic setting.

N10 G90 G0 X0 Y0 Z5 S2000 M3 T1;	Metric basic setting (default state)
N20 G42 G1 X20 Y30 D1;	Metric basic setting (default state)
N30 G1 Z-5. F500.;	Feed at Z
N40 X90	
N50 G20 X2.75 Y3.22;	Positions are entered in inches, G20 effective till deselected with G21 or end of program
N60 X1.18 Y3.54;	
N70 G21 X 20. Y30.;	Positions entered in mm
N80 G0 G40 Z2. M30;	Move out with rapid movement, end of program

Tool Length Offsets (Milling)

Every tool loaded into the machine is a different length (Fig. 3.27). In fact, if a tool is replaced due to wear or breaking, the length of its replacement will likely change because it is almost impossible to set a new tool in the holder in exactly the same place as the old one. The CNC machine needs some way of knowing how far each tool extends from the spindle to the tip. This is accomplished using a Tool Length Offset (TLO) represented by the register "HXX" in the code.

A tool's length affects the position of the tool's tip along the Z axis, and cutting tools vary in length. Tool length offsets (H) compensate for varying tool lengths. Each individual tool requires its own tool length offset to shift the tool tip from the position recorded in the work shift offset to program zero (Fig. 3.28).

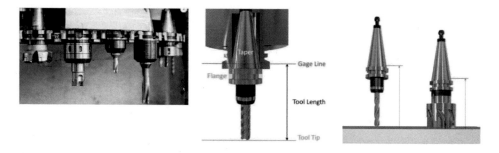

Fig. 3.27 Different length cutting tools installed in an ATC (left). Tool in a spindle with gage line definition (center). Gage line comparison for long versus short tools (right)

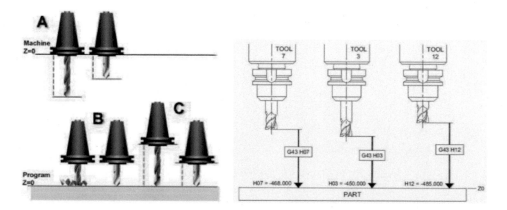

Fig. 3.28 **a** Tools of varying lengths before offset compensation. **b** A tool crash without offset compensation. **c** Tools referenced property with tool length offsets. TLO definition using G43 HXX (Fanuc control, HXX Offset table register)

Tool length offsets can be calculated in a variety of ways. Operators may use different methods to compensate for tool length depending on the convention of the manufacturing facility. Tool lengths are measured with the tool in the toolholder, either on or off the machine. Every tool holder has an imaginary gage line between its taper and flange. Operators measure tool length from the gage line to a tool's tip and then program this value into the MCU. There is a small space between the gage line and the flange that must be accounted for when calculating tool lengths.

When measuring lengths of tools on the machine, an operator touches off each tool against a fixed machine component, such as the worktable. Operators normally place either a piece of paper or a 1-2-3 block between the tool and the worktable, which helps to prevent the tool from damaging the machine (Fig. 3.29). An operator then records the position of the tool, adjusting for the height of the paper or block, and repeats this process for each tool.

The most common way to determine tool length offset is to move the spindle, without a tool, to the point on the workpiece determined by the work shift offset. Then, the operator selects the tool for which tool length offset is to be measured and moves the tip of that tool to the same point on the workpiece. The distance from the gage line to the workpiece surface without a tool in the spindle can be subtracted from the same distance with a tool

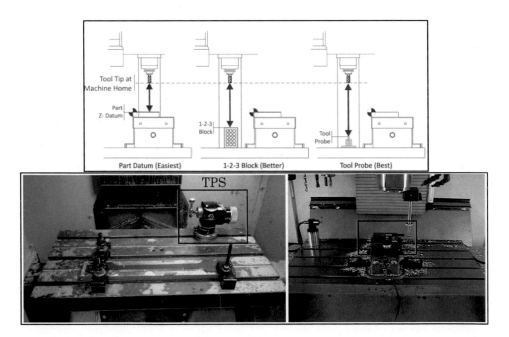

Fig. 3.29 Tool length offsets calculation methods (top). Tool presetting (TPS) probe (bottom left) and Renishaw Ballbar testing for circularity (bottom right)

in the spindle to determine the length of the tool and, therefore, the tool length offset for that tool.

A tool presetting probe (Fig. 3.29) is very similar to the 1-2-3 block method, except the machine uses a special cycle to automatically find the TLO. It does this slowly, lowering the tool until the tip touches the probe and then updates the TLO register. This method is fast, safe and accurate but requires the machine to be equipped with a tool probe. Also, tool probes are expensive so care must be taken to never crash the tool into the probe.

Cutter (Tool) Radius Compensation (CRC)

Most mill controls have cutter radius compensation (CRC). CRC is an offset that adjusts for the radius of the tool, and it is only necessary for tools that travel in the X- or Y-axes. Instead of calculating the toolpath to the center of the tool, the programmer enters a toolpath that follows the contour of the part. The controller compensates for the radius of the tool. The programmer adds either a G41 or G42 to the program to tell the control what direction to compensate for the tool radius. When viewed in the direction of travel, G41 indicates that the tool is moving on the left-hand side of the part, and G42 indicates the tool is on the right-hand side. Most part programs calculate the path of a tool according to its centerline. For tools that travel only vertically on the Z-axis, no additional offsets are required. The reamer and drill shown in Fig. 3.30 are programmed to the center of the tool's tip. Only one tool diameter will match the hole-making operation.

The situation is different for tools that travel along the X-axis and Y-axis. This is especially the case with end mills, which cut with the outer edge. End mills are used to machine pockets, slots, and contour shapes. For these operations, tool diameter affects the dimensions of the part.

Programmers use cutter radius compensation (CRC) to adjust for variations in tool diameter. CRC shifts the cutting tool in a direction perpendicular to its programmed path, as shown in Fig. 3.31. This type of offset enables end mills with two different diameters

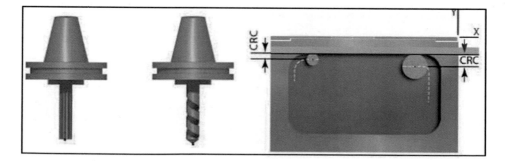

Fig. 3.30 Tools that travel only vertically are programmed to their tip (left). End mills require CRC to compensate for tool (right)

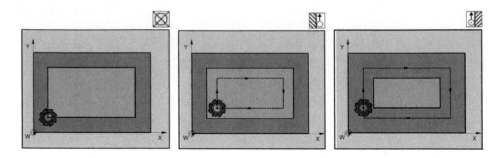

Fig. 3.31 Left: without tool radius compensation (G40, cancel CRC), Center: right-hand tool radius compensation (G42), Right: left-hand tool radius compensation (G41)

to perform the same milling operation. The operator must input the tool radius or diameter for each tool in the CNC controller's offset table (Figs. 3.30 and 3.31).

CRC must be turned on or off with a line move, never an arc. Commanding G40/G41/G42 with an arc move will cause a diameter compensation error that will stop the program. CRC is activated at the end of the line on which it is called, as shown in Fig. 3.32a. Notice how the tool moves at an angle from the start to end of the lead-in line. Programmer needs to activate CRC while the tool is away from the part so this angle move happens away from the finished part surfaces.

As it can be seen in Fig. 3.32a, CRC poses some programming challenges because it requires a ramping motion every time that this compensation is initiated. The ramping motion must be longer than the cutting tool's radius. Depending on the size of the tool, ramping motions can be quite long and cumbersome. The larger the tool, the larger the ramping motion.

To strike a balance, programmers may include G41 codes in the program but enter toolpaths that track the center of the cutter. Figure 3.32b shows a toolpath that factors in the cutter's radius. The operator enters "0.0" as the cutter's radius in the offset table. With this method, the programmer commits to using a particular tool diameter. You must choose a particular tool for each operation, and you can't change it without changing the program. However, CRC becomes essentially a wear offset that the operator can adjust to fine-tune the program. Fortunately, most CAD/CAM software lets you choose how you want to use CRC in the program.

Programmable and Variable Zero Points
Using zero-point shifts, the coordinate origin can be shifted to any desired point within the working area of the NC machine. We can distinguish between:

- Variable zero point shifts: The values by which the zero point is shifted are stored in the zero shift register and are activated by a certain call (e.g.: G54–G59).

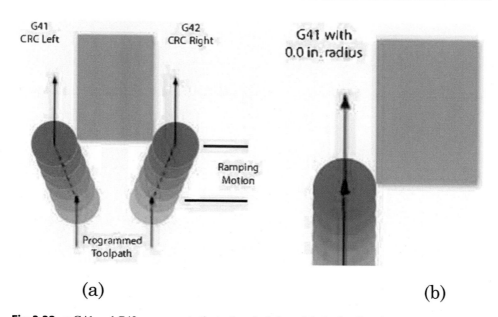

Fig. 3.32 **a** G41 and G42 compensate the tool to the left or right in the direction of the programmed toolpath. A toolpath that factors in the cutter's radius

- Programmable zero-point shifts: The values by which the zero point is shifted are entered and activated directly by a command through the address commands

Example: G59 X10. Y10. Z10.

Zero point shifts (G54–G59)
G codes from G54 to G59 are called zero (work) offsets (zero-point shift). Zero offsets are used to define a new origin with respect to machine zero. Settable zero offsets or adjustable zero offsets. G54, G55, G56, G57 are the settable zero offsets. G58 and G59 are the programmable zero offsets (Fig. 3.29a).

Switching on Zero-Point Shift
In the NC program, the zero point in the machine coordinate system is shifted to the workpiece coordinate system by calling one of the commands G54 to G57. In the next NC block with a programmed movement, all position data and therefore all tool movements are now related to the current workpiece zero point. Using the available zero-point shifts, several workpiece clamping processes (e.g. for multiple processing movements) can be defined simultaneously and called in the program (Fig. 3.33).

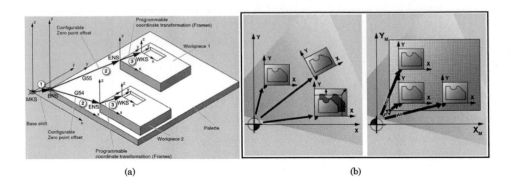

(a)   (b)

Fig. 3.33 Zero-point shifts (G54–G59)

In Practice

Example program

In this example (Fig. 3.27b), 3 workpieces arranged on a pallet according to the zero shift values G54 to G56 are processed one by one.

The processing sequence is programmed in subroutine L47 (not shown).

N10 G0 G90 X10. Y10. F500. T1;	Start
N20 G54 S1000 M3;	Call first zero-point shift, spindle right
N30 L47;	Program runs, in this case as a subroutine
N40 G55 G0 Z200.;	Call second zero-point shift Z over barrier
N50 L47;	Program runs as subroutine
N60 G56;	Call third zero-point shift
N70 L47;	Program runs as subroutine
N80 G53 X200. Y300. M30;	Cancel zero-point shift, end of program

Other Functions

Modern CNC systems have some specially designed functions to simplify the manual programming. However, since most of these functions are system oriented, it is not intended to discuss them here in detail. The following paragraphs give a brief description of commonly used functions in modern CNC systems. The user should refer to the programming manuals of the machine for the detailed programming and operation.

a. *Mirror Image:* This is the function that converts the programmed path to its mirror image, which is identical in dimensions but geometrically opposite about one or two axes.

b. *Program Repetition and Looping:* In actual machining, it is not always possible to machine to the final dimension in one go. This function enables the looping of a portion of the program so that the portion can be executed repeatedly.

c. *Pocketing Cycle:* Pocketing is a common process in machining. This is to excavate the material within a boundary normally in a zigzag path and layer by layer. In a pocketing cycle, the pattern of cutting is predetermined. The user is required to input parameters including the length, width and depth of the pocket, tool path spacing, and layer depth. The CNC system will then automatically work out the tool path.

d. *Drilling, Boring, Reaming and Tapping Canned Cycles:* Canned cycles are routines built into the controller for performing complex operations with a single command (one program block). They save programming time because they can be accessed with a single block of code rather than many individual tool movements. These canned cyles are similar to pocketing cycle. In this function, the drilling pattern is pre-determined by the CNC system. What the user must do is to input the required parameters such as the total depth of the hole, the down feed depth, the relief height and the dwell time at the bottom of the hole.

3.4.1 In-Process Gauging

During many unattended machining operations, such as in manufacturing cells or Agile manufacturing, a periodic checking and adjusting dimensional tolerances of the part is imperative. As the cutting tool wears out, or perhaps because of other causes, the dimensions may fall into the 'out-of-tolerance' zone. Using a touch trigger probe and a suitable program, the In-Process Gauging option offers quite a satisfactory solution. CNC part program for the In-Process Gauging option will contain some quite unique format features. It will be written parametrically and will be using another option of the control system—the *Custom Macros* (sometimes called the *User Macros*), which offer variable and parametric type programming.

If a company or machine shop is a user of the In-Process Gauging option, there are good chances that other control options are also installed and available to the CNC programmer. Some of the most typical options are probing software, tool life management, macros, etc. This technology goes a little too far beyond standard CNC programming, although it is closely related and frequently used (Fig. 3.34).

3.4.2 CNC Machine Calibration with Ballbar and Laser Measurements

The Renishaw QC20-W Ballbar and software is used to measure geometric errors present in a CNC machine tool and detect inaccuracies induced by its controller and servo drive

Fig. 3.34 Touch trigger probe for In-Process Gauging

systems. Errors are measured by instructing the machine tool to 'Perform a Ballbar Test' which will make it scribe a circular arc or circle (Fig. 3.35). Small deviations in the radius of this movement are measured by a transducer and captured by the software. The resultant data is then plotted on the screen or to a printer, to reveal how well the machine performed the test. If the machine had no errors, the plotted data would show a perfect circle. The presence of any errors will distort this circle, for example, by adding peaks along its circumference and possibly making it more elliptical. These deviations from a perfect circle reveal problems and inaccuracies in the numerical control, drive servos and the machine's axes.

In theory, if the user programs a CNC machine to trace out a circular path and the positioning performance of the machine is perfect, then the actual circle would exactly match the programmed circle. In practice many factors in the machine geometry, control system and wear can cause the radius of the test circle and its shape to deviate from the programmed circle. If we can accurately measure the actual circular path and compare it with the programmed path then we would have a measure of the machine's accuracy. This is the basis of all telescopic ballbar testing and of the Renishaw QC20-W wireless ballbar system (Fig. 3.36).

During many unattended machining operations, such as in manufacturing cells or Agile manufacturing, a periodic checking and adjusting dimensional tolerances of the part is imperative.

3.4.3   Principles of Interferometry—The Michelson Interferometer

The use of light interference principles as a measurement tool goes back to the 1880s when Albert Michelson developed interferometry. The Michelson interferometer consists of a light source of a single wavelength (monochromatic), a half-silvered mirror and two mirrors, as shown in Fig. 3.37 below:

Fig. 3.35   Ballbar circular interpolation test

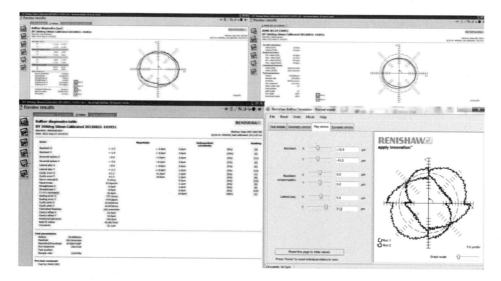

Fig. 3.36   Ballbar circular interpolation test plots for machine health diagnostics

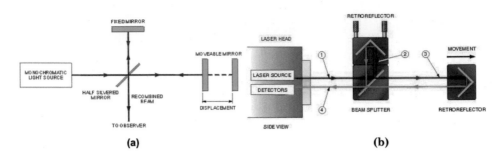

Fig. 3.37 **a** Basic Michelson interferometer **b** Renishaw XL laser system

The light source is split at the surface of the half-silvered mirror, half the light being reflected through 90° towards a fixed distance mirror, the remaining half being allowed to pass through to a moveable mirror. The mirrors are aligned so that the recombined beams reflected from the mirrors are parallel and are reflected back towards an observer. If each of the mirrors is exactly the same distance from the half mirror, then the light will arrive at the observer in phase and constructive interference will occur, resulting in bright light. If the moveable mirror is positioned further away so that its position is shifted by one quarter wavelength, then the beam will return to the observer 180° out of phase and destructive interference will occur, resulting in darkness. Therefore, the distance moved by the moveable mirror can be measured by the observer counting the flashes of light as the mirror moves.

3.4.3.1 Renishaw XL Laser Measurements

Though modern-day interferometers are more sophisticated, measuring distances to accuracies of the order of 1 ppm or better, they still use the basic underlying principles described above. The set-up for a linear distance measurement using the Renishaw XL laser system is shown in Fig. 3.1b. One retro-reflector is rigidly attached to a beam-splitter, to form a fixed length reference arm. The other retro-reflector moves relatively to the beam-splitter and forms the variable length measurement arm.

The laser beam (1) emerging from the XL laser has a single frequency which is very stable with a nominal wavelength of 0.633 µm. When this beam reaches the polarizing beam-splitter it is split into two beams—a reflected beam (2) and a transmitted beam (3). The two beams travel to their retro-reflectors and are then reflected back through the beam-splitter to form an interference beam at the detector, which is housed within the laser head. If the difference in path lengths does not change, the detector sees a steady signal somewhere between the two extremes of constructive and destructive interference [5].

If the difference in path length does change, the detector sees a signal varying between the extremes of constructive and destructive interference each time the path changes. These variations (fringes) are counted and used to compute the change in the difference

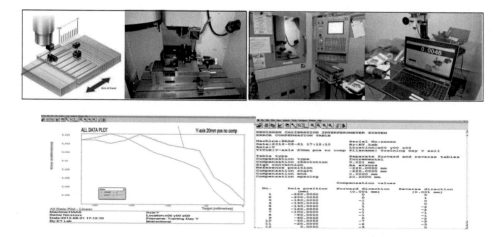

Fig. 3.38 **a** Renishaw XL laser system CNC machine axis linear displacement error measurement. (top) **b** Renishaw XL laser system is being used to measure Haas-OM1 Y-axis linear displacement error (bottom) [5]

between the two path lengths. The length measured will be given by the number of fringes multiplied by the approximate half wavelength of the beam. It should be noted that the wavelength of the laser beam will depend on the refractive index of the air through which it is passing. Since the refractive index of air will vary with temperature, pressure and relative humidity, the wavelength value used to compute the measured values may need to be compensated for changes in these environmental parameters. In practice, for the measurement accuracies quoted in the specification, such compensation is only required for linear displacement (positional accuracy) measurement where the change in the difference between the path lengths of the two beams is significant. Often the intent is to map the errors and then compensate for them in the CNC machine control (Fig. 3.38).

3.5   Introduction to CAM

Once a part design is conceived, it must be entered into the computer. This can be done in a few different ways. In older Computer Aided Design (CAD) processes, a preliminary blueprint is digitized with a special drafting table and drafting device called a puck. The designer moves the puck over the blueprint of the part and clicks it to enter the part's dimensions. Some processes use a scanner to create a raster image file of the drawing. Raster to vector graphics software renders a vector drawing, which can be edited to create a new part. If a prototype is available, its form may be scanned electronically with a laser probe.

In newer and more common CAD processes. Most parts are designed internally using the CAD software itself. Once the preliminary design is in the CAD system, the part model can then be altered as the user sees fit.

Depending on the CAD system, the user can manipulate designs by using commands, choosing from a menu of operations or adjusting the design manually with a computer mouse. When the design of the part is complete, it is ready to be documented and tested.

In 2D modeling, the part model is created as a line drawing. Different colors may be used to distinguish different areas of the part. These drawings are quick and easy to generate for preliminary design work. Although 2D modeling is straightforward and economical, 2D models are inadequate for testing the virtual part. A common type of 2D model is a wireframe model, which uses lines to represent part dimensions. Wireframe modeling can also be used to create 3D images.

3D modeling represents all of the part's surfaces. Designers can manipulate and rotate 3D models to see all of a part's features and dimensions. 3D models can be sectioned to view interior features, shapes and dimensions. 3D modeling may include surface and solid modeling.

Years ago, designing and creating a part meant drawing a mathematically precise image called a part drawing by hand. Today, computers have made this process largely unnecessary. Computer-aided design (CAD) is the use of a computer program to create a representation of a part, including its unique geometry and features. Computer-aided manufacturing (CAM) is the use of computers to define the toolpath instructions needed to create that part. Together. CAD and CAM allow manufacturers to easily manage, control, and streamline all aspects of part creation, including part design, testing, and production.

After the (CAD) process is completed, the first stage in the CAM process is fixture design and cutting tool selection. The part designer determines what tools and work holding devices will be required to make a given part. CAM then interprets the part's geometry and generates the toolpaths directly from the data prepared in CAD. The programmer does not manually encode the tool motion and part geometry, although they may be optimized by the programmer. An engineer may also inspect the toolpaths to make sure there is no interference between tools and other machine components (Fig. 3.39).

Once a part design is conceived, it must be entered into the computer. This can be done in a few different ways. In older CAD processes, a preliminary blueprint is digitized with a special drafting table and drafting device called a puck. The designer moves the puck over the blueprint of the part and clicks it to enter the part's dimensions. Some processes use a scanner to create a raster image file of the drawing. Raster to vector graphics software renders a vector drawing, which can be edited to create a new part. If a prototype is available, its form may be scanned electronically with a laser probe or scanner.

In newer and more common CAD processes. Most parts are designed internally using the CAD software itself (Fig. 3.40). Once the preliminary design is in the CAD system, the part model can then be altered as the user sees fit. Depending on the CAD system, the

Fig. 3.39 Traditional subtractive manufacturing removes material from a piece of stock such as a drill removing metal from a part (left). Additive manufacturing machines work directly from a design created on a computer using CAD software (right)

user can manipulate designs by using commands, choosing from a menu of operations or adjusting the design manually with a computer mouse. When the design of the part is complete, it is ready to be documented and tested.

3D modeling increases accuracy by representing all of the part's surfaces. Designers can manipulate and rotate 3D models to see all of a part's features and dimensions. 3D modeling includes surface modeling and solid modeling (Fig. 3.41). In surface modeling all visible surfaces of the part are shown as plane surfaces, like panels that are fit together to form a hollow object. In solid modeling, a part is modeled as a solid three-dimensional object with interior volume and substance. Many CAM systems require solid models to recognize and auto-generate machining features (feature recognition from 3D-CAD models).

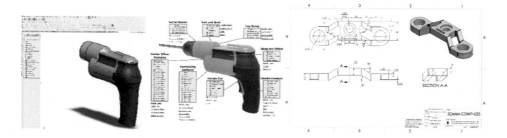

Fig. 3.40   SolidWorks [6] CAD rendering (left) and detailed drawings (right) based on surface or solid models

Fig. 3.41 SolidWorks [6] solid CAD rendering of an engine air intake manifold (left) CNC machined cast parts

In manual preparation of a CNC part program, the programmer is required to define the machine or the tool movement in numerical terms. If the geometry is complicated 3D surfaces cannot be programmed manually. Over the past decades, with the development of the CAD/CAM systems, interactive graphic systems are integrated with the CNC part programming. Graphic based software using icon driven techniques improves the user friendliness. The part programmer can create the geometrical model in the CAM package or directly extract the geometrical model from the CAD/CAM database. Built-in tool motion commands can assist the part programmer to calculate the tool paths automatically. The part programmer can verify the tool paths through the graphic display using the simulation function of the CAM system. It greatly enhances the speed and accuracy in tool path generation (Fig. 3.43).

3.5.1 Workflow of a CAM System

There are many CAM or CAD/CAM systems available in the market. The general workflow to go from CAD model to machined CNC part is outlined below:

1. Begin with CAD modeling.
2. Establish the job parameters including CNC coordinate system and stock shape/size.
3. Select the CNC process.
4. Select the driving CAD geometry (Fig. 3.42, left and middle).
5. Select the cutting tool and machining parameters (Fig. 3.42, right).
6. Verify toolpath (Figs. 3.43 and 3.44, left).
7. Post Process (Fig. 3.44, right)
8. Transfer G--code program to CNC machine. (via data transmission)
9. Set up and operate the CNC machine to make part.

Fig. 3.42 Automatic feature recognition based on imported solid models Left: SolidWorks CAM Center: Autodesk FeatureCAM [7] Right: CAM Works [6] mill Tool set-up

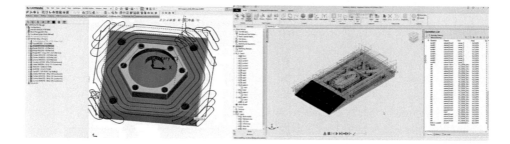

Fig. 3.43   Tool path generation Left: CAMWorks [6] Right: Autodesk Feature CAM [7]

Fig. 3.44   Tool path simulation (Left: FeatureCAM [7]), G-Code verification and potential collision detection with VeriCUT Simulation (right) [9]

After the CAD process is completed, the second stage in the CAM process is fixture design and cutting tool selection (Figs. 3.42 and 3.44). The part designer determines what tools and work holding devices will be required to make a given part. CAM then interprets the part's geometry and generates the toolpaths directly from the data prepared in CAD (Fig. 3.42). The programmer does not manually encode the tool motion and part geometry, although they may be optimized by the programmer. An engineer may also inspect and verify the toolpaths to make sure there is no interference between tools and other machine components such as fixtures and clamping devices (Fig. 3.44).

The verification methods included within CAM systems may be inadequate for complete and accurate simulation and verification of G and M codes (the CNC program). A CAM use should be able to catch material crashes, gouges, and watch the material being removed to make CAM operation improvements. Typically, the CAM system's verification does not track data such as excessive cutting forces, deflection, or recognize improvement opportunities for faster feeds, which results in varying chip thicknesses.

A more robust verification and optimization system is needed to identify these issues and make these changes (Fig. 3.44, right). It is generally difficult to know if a cutting tool (milling solid tool, indexable insert, or turning insert) is being used to its full engineered potential. CNC programmers needs to verify that their tools will not suffer adverse effects like excessive feeds, incorrect chip thicknesses (too thin or too thick), excessive cutting forces, or tool deflection [8].

CNC simulation software such as VeriCUT, provides detailed cutting condition graphs that can provide visibility to the issues. The ability to have graphs in combination with the 3D digital twin graphics give a CNC programmer powerful visual feedback to aid in making program improvements. A programmer can see a simulation of the cutting tool in the material. Along with the position in the CNC program and graph showing the chip thickness, force and tool deflection at any point (Fig. 3.45). Graphs can also provide insight into torque, power, material removal rate, and other useful data that a CNC programmer can use to make intelligent machining decisions.

Review Questions

1. What is the main difference between NC and CNC machines? Briefly elaborate.
2. What makes a CNC mill a CNC machining center?
3. List at least four advantages of CNC technology over manual machining.
4. How do you designate positive axes directions for?

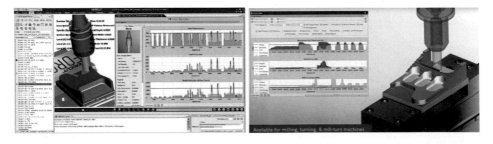

Fig. 3.45 NC program, digital twin, graphs for chip thickness, total force, tool deflection shown in VERICUT simulation [9].

    a. A vertical CNC machining center?

    b. A horizontal CNC machining center?

5. Define the terms "machine zero, work zero, work offset, and part/program zero".

6. What are the axes for a 5-axis CNC machining center? Explain each one briefly.

7. What is the main functional difference between G & M codes?

8. Give the names of two different positioning (programming) systems and explain the difference between them? What G-code do you use to differentiate between them?

9. What is the difference between modal and non-modal G & M codes?

10. Name three miscellaneous (M) codes and their functions?

11. What is a "fixed or canned cycle"?

12. How does a CNC lathe code vary from a CNC mill code? Use an example to explain.

13. Define the terms "tool post, and turret" used for CNC lathes and turning centers?

14. What are the three types of inputs needed to define a cutting operation in a CAM package?

15. Define the phrases "backplotting/playbacking a tool path, verifying a toolpath, and post-processing".

Research Questions/Challenges:

1. Conduct a review for understanding high speed milling machines and their uses. Summarize your findings in two PowerPoint slides.

2. What does the term "qualifying a cutting tool" mean? Explain in a concise manner.

3. Research how Haas VF1 or a similar CNC machine with a tool probe qualifies its tools. It is a different procedure than what is described in this chapter.

4. Haas VF1 or a similar CNC machine with a probe allows setting of its work zero (or program/part zero) manually or automatically. Please research and summarize each procedure concisely with sketches.

5. What are the common materials used in cutting tools? Prepare a mind map summarizing them and their uses.

6. What is the difference between "climb" and "conventional milling"?

7. What does the unit "sfpm" represent? (Hint: the unit is related to the S code representing spindle speed in rpm)

8. What does the term "flute" mean? How does it impact the cutting operation?

9. Find a G & M code example for each of the following NC content: mirror image, program repetition and looping, pocketing cycles, drilling, boring, reaming and tapping canned cycles.

10. Find a brief NC (G & M) code, possibly 2D, and draw the geometry it will produce when it is run on a graphing paper.

11. Write a manual CNC program that will machine the outline in CCW direction. Program should start and end at tool change/set position. Axial depth of cut 0.375 in., cutter diameter 0.5 in., 2 flutes end mill. Assume stock is oversized with respect to drawing

and you are just machining the geometry in the drawing (not scaled). Assume material being machined is nylon. Using https://ncviewer.com/, simulate and verify your G&M code.

12. How can a package like Mastercam engrave a photograph (given in.JPEG or PNG)? Explain the software resources needed to complete the task?

13. Research NC expert machinist systems via their process, advantages, and disadvantages. (Hint: Look into Creo NC Expert Machinist and Esprit software packages)

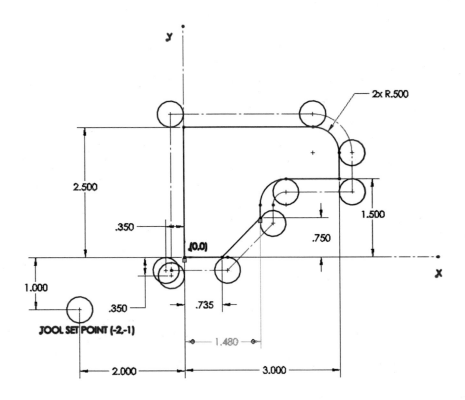

14. Using a CAD/CAM system of choice, generate tool paths for the part in the detail drawing given below.

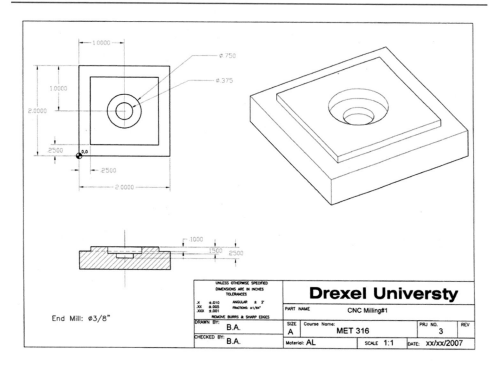

End Mill: ⌀3/8"

UNLESS OTHERWISE SPECIFIED DIMENSIONS ARE IN INCHES TOLERANCES		**Drexel Universty**			
.X ±.010 ANGULAR ± 2° .XX ±.005 FRACTIONS ±1/84° .XXX ±.001					
		PART NAME	CNC Milling#1		
DRAWN BY: B.A.		SIZE Course Name:		PRJ NO.	REV
CHECKED BY: B.A.		A MET 316		3	
		Material: AL	SCALE 1:1	DATE: xx/xx/2007	

15. Using a CAD/CAM system of choice, generate tool paths for the bottle opener in the detail drawing. Use the tool and process plan data in the attached tool and production sheet.

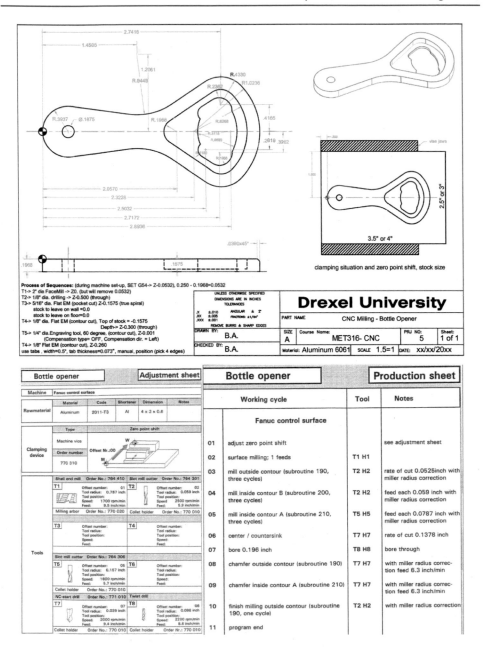

16. As the CNC lathe turns down the diameter of the stock, the cutting speed is supposed to decrease due to the stock losing its diameter. How can that be remedied?
17. Find examples of turning operations available in a CAM program, identify the processes and the tools used/

18. Conduct a literature review on generating machining or turning process cost analysis. Attach an example case.

Discussion Questions

1. Why work holding is an important part of manual and NC/CNC machining. Elaborate on it thoroughly.
2. Machining a hardened material may often become a necessity. How do you handle it? What type of machines and cutting tools can be used in machining a hardened die or mold?
3. Discuss the current role of CNC machining in rapid prototyping, tooling, and manufacturing.
4. Discuss the role of wire and sinker EDM in tool making.
5. Discuss the future of CNC machining as it also relates to AM.

References

1. Groover, M. P. (2003). *Automation, Production Systems, and Computer-Integrated Manufacturing 2nd Edition*, Upper Saddle River, NJ: Prentice Hall. p. 8.
2. ISO Standards Handbook 7, 1981.
3. Emco Meijer Concept Turn 250 Programming Manual.
4. ANSI/EIA Standard, RS-274-D.
5. Methods for Performance Evaluation of CNC Machining Centers, ASME B5.54–1992.
6. Dassault Solidworks- CamWorks Mill tutorial, 2022.
7. AutoDesk, FeatureCAM Training Manual.
8. Haas, P. (2021), True Constant Chip Thickness Machining, The New Standard of NC Program, White Paper, CGTECH, 2021.
9. Cgtech VeriCUT Training Manual.

3D Printing Pre-processing

4

4.1 Introduction to 3D Printing and AM

3D Printing or AM is a process of making three dimensional objects from a digital geometry. Various names are used for different technologies, but most are based on the process of creating objects layer-by-layer. The object is created in different fashions including extruding and laying down successive layers of softened solids, curing layers of liquid materials, sintering or melting powders in layers. Each of these layers can be seen as a thinly sliced horizontal cross-section of the final object. With 3D printing and AM, ideas come to life. In just a few hours, users can have a prototype model that can be examined from every angle. The use of 3D printing can rapidly produce a high-quality prototype model, called a rapid prototype for concept development and functional prototyping. With Rapid Prototyping (RP) engineers can communicate their design concepts and have more time and opportunities to perfect it before production. Another application is employing 3D Printing in the manufacturing phase, where the end-use product is produced (AM) (Fig. 4.1). A facet of digital manufacturing, called rapid tooling (RT), can be utilized for tooling, jigs and fixtures in addition to creating production parts [1, 2] (Fig. 4.1c). An in-house 3D printer can be the owner's personal factory: producing the product, be it jigs and fixtures, molds and patterns, or small volume or on-of-end-use parts. That's having production without a line, reducing delay and expenses of tooling and inventory, producing on-demand, customized, geometrically complex products, free from traditional manufacturing constraints.

The field of AM developed in many different locations at the same time. Independent engineers and companies all over the world created new methods and prototyping techniques which eventually became what is known as AM. It today includes methods as different as fused deposition modeling (FDM), stereolithography (SLA/SL), and sheet lamination. Despite their wide range of variety, the process (workflow) used in most of

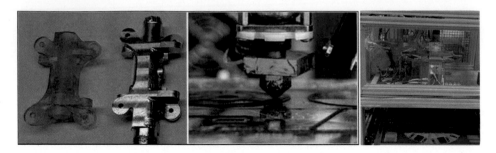

Fig. 4.1   a AM (right) has evolved from techniques used to make prototypes (left) **b** FDM technology can be used in making custom end-use parts including prosthetic sockets **c** AM fabricated test bench for automated testing of Injection Molded Parts

these AM methods is largely the same. Most AM operations follow the same workflow [3]:

1. Design the product in a CAD package or scan a pre-existing object into CAD.
2. Convert the CAD geometry to an intermediate file format like stereolithography (.STL file).
3. Convert the intermediate file format into a build file (i.e. .gcode).
4. Set up the AM machine.
5. Transfer the build file to the AM machine.
6. Run the machine to build the product.
7. Retrieve the product and remove any support materials or support structures from the part. (Some AM processes may not need support structures)
8. Perform any necessary post-processing on the product.
9. Use the product for the final application.

4.2   Designing in CAD

Every finished product begins as a design. A key advantage of AM is that the product is directly built by a machine working from the design itself. As a result, not only are any necessary changes to the final product's design easily accomplished but moving to the actual production phase also occurs much sooner in AM than other traditional manufacturing processes. To work directly, all products that are manufactured by AM are originally designed in a CAD package.

Product design can be handled in several different ways in a CAD program. Most often, an individual creates the solid model and consequently its blueprint for the product directly inside the CAD. The user then assembles the exterior boundary of the

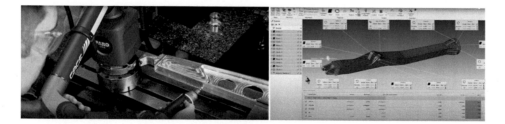

Fig. 4.2 Reverse engineering equipment FARO 6-axis inspection arm allows a person to recreate a product geometry in CAD using an already manufactured object

design using polygons by converting the CAD model into the STL file. However, reverse engineering (RE) equipment, such as a coordinate-measuring machine (CMM) or laser scanners that determines an existing product's dimensions using a probe, is also becoming more commonly used during this process. Reverse engineering equipment allows a person to capture three-dimensional geometric data from an already manufactured object to use as a design. Though this technology continues to improve, almost all product blueprints generated through reverse engineering will still require a person to make minor changes and corrections before using the design for AM (Fig. 4.2).

When using CAD software for design, check for any inconsistencies caused by shapes that are not fully enclosed. Although most modern CAD software will do this automatically, an additional check can identify any remaining gaps in the product design and ensure the image is watertight in order to prevent major flaws in the final product. The lead author employed Geomagic's Studio in the past in patching the holes in a surface model (see Sect. 9.2.2.7).

4.3   Design for 3D Printing and Additive Manufacturing (DfAM)

When an engineer designs a part or a system in a CAD program, it will always look perfect on the computer screen. However, what it will look like as a physical object will depend on how it is manufactured. It's always a good idea to remember that a CAD rendering is a work of art that describes an imaginary object that does not (yet) exist. Different manufacturing techniques and materials might make your part's dimensions change in one axis and not another, or perhaps the part can't be made at all by some technologies. Often, some changes will need to be made to have the part efficient enough for manufacture. The process of optimizing a design so that it is best for a given manufacturing technique is called design for manufacturing (DFM).

Design for additive manufacturing (DfAM), which is what we call the specific skills involved in creating parts and assemblies that can be additively manufactured smoothly and with a minimum of manual postprocessing, and associated costs [4]. Hopefully, these

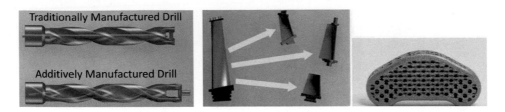

Fig. 4.3 Additively versus traditionally manufactured drill (left), Customized turbine blades with internal cooling channels (right)

ideas will increase the probability that the finished part or assembly will function the way CAD design intended. From Fig. 4.3, one can see how much simpler an additive design can be compared to a traditionally manufactured one. In this example, manufacturers use AM to create tooling for traditional manufacturing. AM processes not only create tooling quickly, but AM tools can also be created as a single, unified part, eliminating, or reducing the need for assembly. Manufacturers have used direct metal laser sintering (DMLS) to create drills with optimized internal channels that allow for improved cutting fluid flow (Fig. 4.3, left). Improved flow enhances drill performance and lengthens drill service life. By printing this part on the DMLS printer, the tooling can be produced faster and more affordably than traditional machining.

The most widespread, and growing, use of AM is to create end-use, near net-shape parts. End-use parts are AM components that are used directly by a general consumer, such as footwear or sporting accessories, or components that industrial manufacturers will add to larger assemblies, such as automobile drive shafts or jet engine turbine blades (Fig. 4.3, center, right).

4.3.1   AM Design Constraints

Additively manufactured parts are mostly built up a layer at a time. DfAM is a design methodology recommended for a specific additive manufacturing technology. DfAM incorporates the unique building characteristics of additive manufacturing into the design of a component to be produced using AM [5].

There are numerous design modifications and considerations that have an impact on the printing time, material consumption including supports, part's strength, as well as parts with added value to various manufacturing processes. In addition to selecting the best printing orientation (*Build Orientation*), here are a few key considerations when designing a part for Material Jetting (such as Stratasys PolyJet) printers (Fig. 4.4, Table 4.1):

- Consider the part's strength requirements and modify the design accordingly.

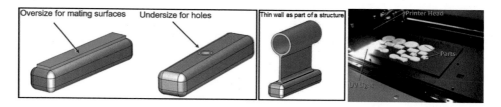

Fig. 4.4 When high tolerances are required, mating surfaces can be designed oversized and holes should be undersized, and subsequently machined (light blue area) to desired tolerances (left). The wall is part of a structure or a larger model, the thickness of the wall should not be less than 1 mm (0.03937 in) (center). Parts designed for PolyJet printers (right) should have a minimum wall thickness of 0.6 mm (0.024 in)

Table 4.1 Geometries that can cause fabrication issues

Technology	Problematic geometries
Filament	Overhangs
	Feature sizes < about 1 mm
	Tall, thin prints (particularly ones that come to a point)
	Walls, thin fins <1 mm thick
Liquid resin	Overhangs
	High aspect ratio, small features
	Internal voids or cups that pool liquid resin Large solid prints
Bound metal deposition	Mostly same as filament (for details, see such as Desktop Metal "Bound Metal Deposition")
Metal binder jetting	Internal fully enclosed voids that hold unsintered powder Small gaps between print-in-place parts
	Parts that need support will require equivalent during sintering
Direct sintering	Internal fully enclosed voids that trap powder
	For metals, prints requiring significant support (warping)

- Use minimum wall thicknesses when possible.
- Eliminate not needed overhangs to decrease the required support material.
- Print using different fill styles (*Fill Patterns and Lattice Structures*), keeping the part's durability, while reducing material consumption and printing time.
- Print assemblies in one print run, to minimize assembly labor time and costs, and reduce the chance of human error during the assembly process.
- Consolidate an assembly of parts into one, to minimize part count and post-processing (*Part Consolidation*), as well as improve part reliability.

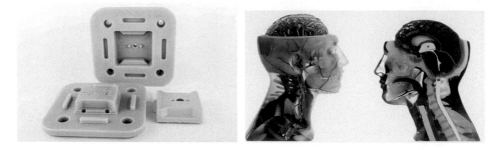

Fig. 4.5 Low-run injection mold printed from Digital ABS, used to produce a batch of sensor housings. (Image courtesy of Protomolding) (left). The medical industry often utilizes full color printing to produce educational medical models. (Image courtesy of Stratasys) (right)

As with most 3D printing technologies, material jetting requires the use of support material to accurately print parts. Like some FFF printers, material jetting prints support from a secondary dissolvable wax-like material. The need to manually remove support material places limitations on the design of material jetting parts. Any fully enclosed cavities will be filled completely with support material that cannot be removed. Any holes or channels should be greater than 0.5 mm in width (channels with a depth-to-width ratio of 2:1 are especially difficult to clean). Escape holes generally do not assist with the removal of support material and are therefore not required, as the support material is printed as a solid (compared to SLA or SLS where the material being removed is a liquid or powder) (Fig. 4.5).

Wall Thickness
When dealing with wall thickness, we need first to understand slice height and toolpath's width, and how they can affect wall thickness (Fig. 4.6). The slice height is the distance from the tip (FFF/FDM technology) to the previous layer. The width of the tool path (*road*) is determined by the speed of the drive wheels, which regulates the amount of extruded material, the faster the drive wheels movement, the more material is extruded. More material means wider toolpath and less material will generate a thinner tool path.

The minimum wall thickness for FDM parts depends on slice thickness that will be used to build the part. The thickness of vertical walls should be at least twice the slice thickness in order to avoid creating walls that are prone to buckle. Making walls four times the slice thickness or thicker will provide even stronger walls. Another consideration is a vertical wall should be composed of at least two filaments so wall thickness should be at lcast twice the extrusion path (Fig. 4.7).

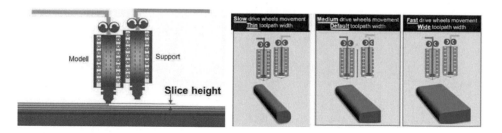

Fig. 4.6 The slice height and tool paths in FDM/FFF. Each wall/rib needs to be printed with at least two roads. Minimum wall thickness is two times the toolpath width. If designing thinner walls there will be an overfill. The walls will be thicker than the design and part quality can get worse

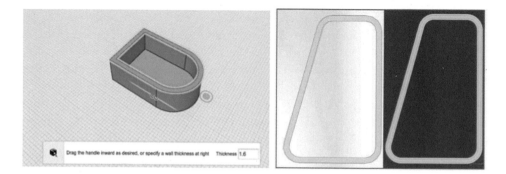

Fig. 4.7 If you are making boxes or enclosures, you should make sure that the walls are an exact even multiple of the layer width (left). The reason for this is that the slicer has no way of accounting for very small gaps and will simply not extrude anything. The wall thickness, as well as inner and outer fillets should be constant (right)

Overlapping Geometry and Boolean Operations

GrabCAD Print or similar slicing software can struggle to interpret the design intent when encountering overlapping geometry. To avoid these occurrences, the sections should have the appropriate Boolean operations performed so as to create either single bodies or defined separate bodies where different material designations are required (Figs. 4.4 and 4.8).

Designing Assemblies for All-in-One Manufacturing

Integrated Design (Part Consolidation)

If reproducing an existing part, one can integrate as many components as possible into one piece. If designing a new item, one can also create it as one piece [4]. It is recommended to only split off parts when it is advantageous to the part's operation or maintenance. Integrated design has many advantages including:

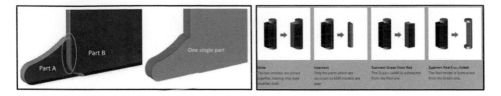

Fig. 4.8 Overlapping in the joining areas (left), Boolean operations (right)

- *Improved functionality:* Focus on the task that the part will perform, optimizing its design for function, rather than the process used to make it.
- *Eliminate tolerance challenges:* Holding tight tolerances is costly. If two mating parts are combined into one, concerns related to controlling the tolerances are eliminated.
- *Eliminate assembly time:* Assemblies, obviously, must be assembled and this takes time. Consolidating all parts into a single piece eliminates the time needed for assembly.
- *Decrease bill of materials (BOM) count:* Smaller part counts decrease the time and expense for managing and warehousing inventory.

With AM, it is possible to print moving part assemblies, pre-assembled in the build tray (Fig. 4.9). How do we make sure production of moving parts and assemblies without getting stuck or jammed?

Moving Parts
Clearance between the moving parts need to be considered. It is recommended to use a clearance of 0.2–0.4 mm (0.008–0.02 in.) between moving parts. The clearance areas are both in the vertical and horizontal planes. The clearance in the vertical plane (horizontal clearance) will be printed with support structure (Fig. 4.10).

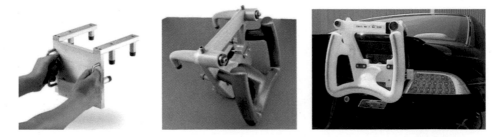

Fig. 4.9 From left to right 1. Multi-piece assembly of a fixture produced using traditional manufacturing. 2. 3D printed consolidated part 3. Part used on the car

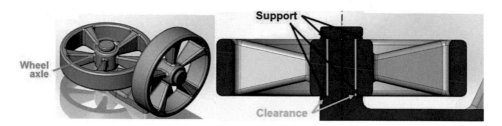

Fig. 4.10 Support generation for moving assemblies

Assembled Parts

Components are often dissected into many pieces to make conventional manufacturing processes feasible and affordable. This is unnecessary with typical AM process such as FDM. Assemblies can be converted into single parts. If that's the case, a clearance is necessary for sliding or moving parts (Fig. 4.11). The standard guideline for creating clearances on assemblies being produced fully assembled: A minimum Z clearance of the slice thickness. The X/Y clearance is at least the default extrusion width based on a suggested minimum wall thickness. Although it might be best to print parts separately, if they are printed assembled, a clearance should be added as follows: In the XY axes— clearance of 0.2 mm, in the Z axis—clearance of one slice height, in-order to close without a visible gap. For good fitting parts printed separately to be subsequently assembled will require 0.05–0.1 mm clearances between moving parts (Fig. 4.11, top row).

Rotating Assemblies

Create slots or openings in rotating assemblies to help remove support material. For assembled parts printed separately, the minimum clearance needed for mating parts, when

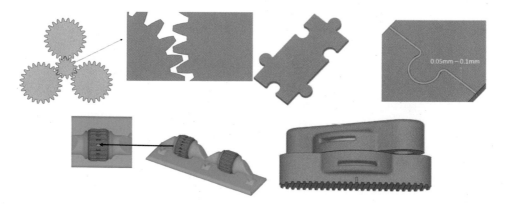

Fig. 4.11 There are differing requirements for moving and assembled parts and rotating assemblies

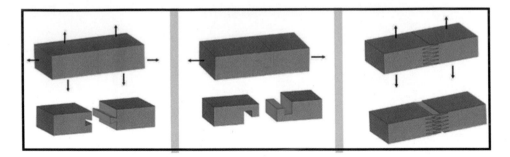

Fig. 4.12 This sectioning type is good for pulling forces applied from the sides as well from above and below (left). This sectioning type is good for pulling forces applied from the sides (center). This sectioning type is good for pulling forces applied from above and below (right)

not producing the components fully assembled, is equal to the tolerance of the FDM machine itself (Fig. 4.11, bottom row).

STL File Sectioning
Sectioning is especially useful for parts that are larger than the build volume of the printer. Some CAD packages offer a tool that enables the user to select a plane through the model and have it sectioned automatically, then insert location features like dowels or other geometry, besides employing cold welding in simpler parts. This can also be achieved manually. These features improve the strength of the finished part and help to ensure that the parts line up correctly. The main features to consider when creating and orienting interlocking joints include the friction, tension, and shear forces involved. Here are some of the questions that need to be asked (Fig. 4.12):

- How tight does the joint need to be?
- Does it need to be assembled with force or with ease for regular assembly and disassembly?
- Does it need room for the adhesive to be effective?
- Should the sections hold together by themselves with a friction fit?

There are different joint options that can be used, such as dovetail joint, half lap joint, mortise joint and combination of joints.

Overhangs and Support Structures
Overhangs are one of the best-known constraints in FFF printers. Since the layer width is greater than the layer height, the deposited filament has a rectangular section, and overhangs are possible. Generally, the plastic is "sticky" enough to easily allow overhangs of up to 45° off the vertical. However, you can go even further, up to 70°, if you have a narrow salient. The side walls are supported well, so you are in effect "*bridging*" across

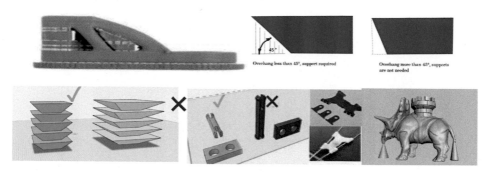

Fig. 4.13 Overhangs should be minimized (top). Ideally, parts are oriented in such a way so that the greatest stress is carried along the length of the filament—the orange parts illustrate this (bottom). The red parts would be a lot weaker. Create multiple parts, so that all are strong (bottom, center)

the bottom surface. If this is not the case, the 45-degree rule applies. In Fig. 4.13 (bottom left) the green tower will print fine, but not the yellow one—the extruder would be printing into air. Of course, support material is available and works well in this case—however, it will increase print time and affect surface finish. It is preferred to design overhangs that are connected to the main object, even when using support. Additional helpers can be used to stabilize disconnected overhangs. In some cases, you'll have overhangs that are disconnected from the main body of an object. In these cases, you want to design in support for them, even if you will later use the slicers own support generation tools.

The support generated by the slicer is intentionally "flimsy" and won't properly hold up structures like the nose and tail of the dinosaur shown, so you'll want to give them a stable base yourself (Fig. 4.13). Using support material is a great enabler for complex models, however you should use it judiciously [5, 6]. Parts need to be oriented to minimize the amount of scaffolding needed (Fig. 4.13), and so that the support attaches to surfaces that are not visible.

FFF printers are capable of bridging quite large gaps if both ends are supported (Fig. 4.14). 50 mm is certainly possible, and for some printers even greater distances can be accomplished. This is done by reducing the extrusion rate of "bridges" so that the filament is slightly stretched. However, for this to work, you have to make sure that the spans are completely flat—otherwise you're dealing with an extreme overhang situation and need to use support.

Mechanical Properties and Build Orientation
Mechanical properties are best in the X–Y build plane and weaker across layers (FDM/FFF). Multiple loading conditions may require an orientation that provides the best compromise for each case. An important limitation of Fused Deposition Modeling (FDM) printing is the anisotropic nature of the parts that it produces. Anisotropic materials have varying mechanical properties in different directions. Timber is a good example

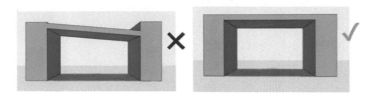

Fig. 4.14 Use of bridging: level spans anchored at both ends can be bridged without support—50 mm is certainly possible (right). Make sure that unsupported spans are flat to allow bridging to work (left)

of an anisotropic material. It is often the adhesion between the layers that defines the strength of an FDM part rather than the material it is made of. Adhesion of layers is dependent on printer calibration and settings and is the responsibility of the operator. Due to the anisotropic nature of FDM printing, understanding the application of a component and how it is built are critical to the success of a design.

FDM components are inherently weaker in one direction due to layer orientation (Fig. 4.15). With the Fused Filament Fabrication (FFF) printing process, layers are pressed down upon one. Since the layers are printed as rounded rectangles, between each layer there are small valleys. These valleys create stress concentrations where a crack may originate when the part is placed under load (Fig. 4.16).

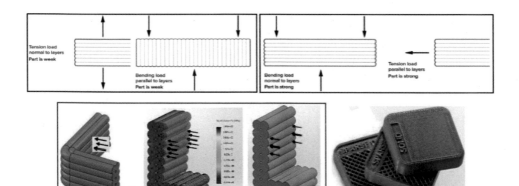

Fig. 4.15 Anisotropy, or the weakness between layers, will cause a part to react differently under different forces (top row [14]). Although FFF is regularly defined as the simplest 3D printing technology, most of design limitations and rules center around the anisotropic behavior of FFF parts and the need for support material. The left part in (bottom row)) is printed in Z orientation whereas the right part is printed in XY orientation. Color plot in part **b** indicates FEA simulation showing YZ shear stress in this printed orientation when loaded in horizontal X-force (cyan color). Infill geometries for FDM (blue parts, bottom-right) (image courtesy of Stratasys [7]

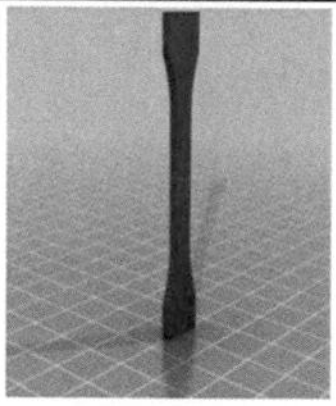

	Printed vertical (Z-axis)		Printed horizontal (X-Y -axis)	
Infill	**50%**	**100%**	**50%**	**100%**
Tensile Strength (MPa)	4.4 ± 0.6	6.5 ± 1.8	17.0 ± 0.8	29.3 ± 0.8
Force at break (MPa)	2.7 ± 1.8	7.8 ± 1.3	13.6 ± 0.8	26.4 ± 1.8
Elongation at max. force (%)	0.5 ± 0.1	0.7 ± 0.1	2.3 ± 0.1	2.4 ± 0.1

Fig. 4.16 An FFF datasheet from ABS showing the mechanical properties of a test specimen printed in 2 different orientation and infill percentages (top) [8]

Let's assume we are designing two brackets (Fig. 4.15, left). One of the brackets would be weak in the direction of the force, whereas the other bracket would be substantially stronger because of the print orientation. When designing your component, it is good to understand the orientation that it will be built in, so that it can be designed with the optimal number of features that will be subjected to high forces oriented in the X–Y plane.

Infills, Lattice Structure, and Topology Optimization
Infill styles can also be changed according to the part's requirements (Figs. 4.17 and 4.18). Solid, sparse double density/sparse high density, and sparse low density build styles range from the production of solid and durable parts to the production of light weight, and cost and time-efficient parts, and draft printing mode for faster and economical printing options. The different build styles are part of the advanced options of the build-preparation programs, giving additional and important control over the production parts, time, and costs.

In addition to the infill styles, new software packages like nTopology offer *lattice structures* that resemble 3D cells that make up crystalline materials as well as shape optimization for minimizing mass of the part without hurting its function (Topology Optimization as illustrated in Fig. 4.19.

It goes without saying that design mesh structure should be in good shape. If one is using CAD programs that are solid-aware, it will probably generate a good mesh. However, it is worth to consider the following (Fig. 4.20):

- Solids: Avoid surfaces without any thickness,

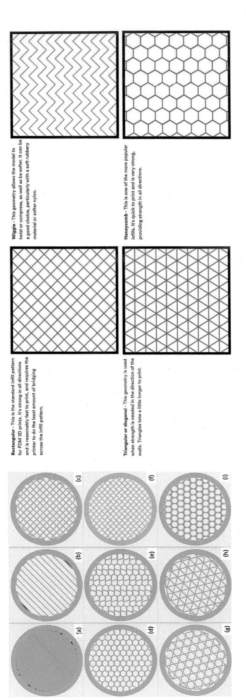

Fig. 4.17 Standard infill geometries (left [7] and right [8]) **a** Solid—Best strength, **b** Sparse—Fastest print, least amount of material, poor strength, **c** Sparse Double Dense—Fast print with good strength, **d** Hexagonal—Best combination of strength and material used, **e** Hexagonal Porous—For circulation between cells, **f** Sawtooth, **g** Hexagram—Soluble tooling, **h** Permeable triangular—Soluble tooling, **i** Permeable tubular, Soluble tooling. (image courtesy of Stratasys [7] and Hubs [8])

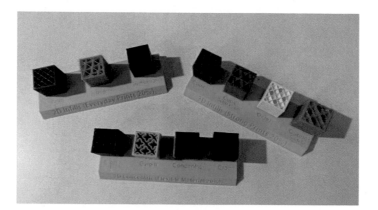

Fig. 4.18 Infill pattern samples printed at 20% infill rates

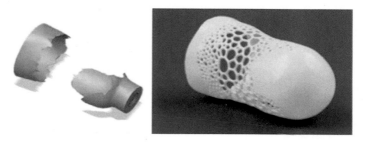

Fig. 4.19 Topology optimization used in FFF printed prosthetic sockets a. preserved areas b. resulting socket after topology optimization (completed at RMU) [9]

Fig. 4.20 Good CAD oriented programs should generate good quality error free meshes (left)—however, if you are using surface modelers, or working with imported meshes, then you should use analysis and repair tools before printing (Meshmixer-right) [10]

- Watertight: There shouldn't be any gaps or cracks—otherwise it will be treated as a hollow shell.
- Normals: Make sure that surfaces (normals) are correctly oriented facing out.

	Supported walls	Unsupported walls	Support & overhangs	Embossed & engraved details	Horizontal bridges	Holes	Connecting /moving parts	Escape holes	Minimum features	Pin diameter	Tolerance
	Walls that are connected to the rest of the print on at least two sides.	Unsupported walls are connected to the rest of the print on less than two sides.	The maximum angle a wall can be printed at without requiring support.	Features on the model that are raised or recessed below the model surface.	The span a technology can print without the need for support.	The minimum diameter a technology can successfully print a hole.	The recommended clearance between two moving or connecting parts.	The minimum diameter of escape holes to allow for the removal of build material.	The recommended minimum size of a feature to ensure it will not fail to print.	The minimum diameter a pin can be printed at.	The expected tolerance (dimensional accuracy) of a specific technology.
Fused deposition modeling	0.8 mm	0.8 mm	45°	0.6 mm wide & 2 mm high	10 mm	Ø2 mm	0.5 mm		2 mm	3 mm	±0.5% (lower limit ±0.5 mm)
Stereo-lithography	0.5 mm	1 mm	support always required	0.4 mm wide & high		Ø0.5 mm	0.5 mm	4 mm	0.2 mm	0.5 mm	±0.5% (lower limit ±0.15 mm)
Selective laser sintering	0.7 mm			1 mm wide & high		Ø1.5 mm	0.3 mm for moving parts & 0.1 mm for connections	5 mm	0.8 mm	0.8 mm	±0.3% (lower limit ±0.3 mm)
Material jetting	1 mm	1 mm	support always required	0.5 mm wide & high		Ø0.5 mm	0.2 mm		0.5 mm	0.5 mm	±0.1 mm
Binder jetting	2 mm	3 mm		0.5 mm wide & high		Ø1.5 mm		5 mm	2 mm	2 mm	±0.2 mm for metal & ±0.3 mm for sand
Direct metal Laser sintering	0.4 mm	0.5 mm	support always required	0.1 mm wide & high	2 mm	Ø1.5 mm		5 mm	0.6 mm	1 mm	±0.1 mm

Fig. 4.21 Comprehensive list of DfAM rules (Image courtesy of hubs.com) [8]

- No self-intersections: This will throw off the odd/even calculation and lead to strange voids (Fig. 4.20)
- Use Boolean unions: even if your model looks correct, make sure to use whatever your software allows for to make the model into a single solid—otherwise you'll get strange (and weak) results.

The following table in Fig. 4.21 presents an overall summary of all of the design rules, per technology, presented in this section. This table serves as a useful reference when looking to determine which technology is best suited for a particular design. It should be noted that the values presented in this table are general recommendations and may vary based on specific printer type or the material used. [8]

4.4 Intermediate File Format Generation

CAD programs contain a great amount of extra data that is not necessary for additive manufacturing. Most CAD programs track the history of the modeling done on the design, as well as other extraneous information that is useful during the design process but not needed for the actual manufacturing of the finished product. Thus, most AM machines will not be able to work directly from a CAD file. Instead, nearly every AM machine uses the.STL file format for the process. The *Standard Tessellation Language/Stereolithography* (STL) file format is the pioneer 3D printing file format. It was invented in 1987 by Chuck

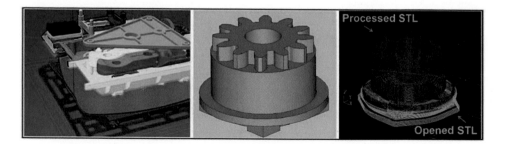

Fig. 4.22 While an AM product is initially designed with a CAD program, a CAD file contains unnecessary data (left). Most AM machines use STL file formats to build a product rather than the CAD files with which the product was designed (center). Sliced STL file in Catalyst (Courtesy—Stratasys) (right) [7]

Hull [11], the inventor of the SLA process. The STL file format is still the most widely used and is considered to be the standard file format in 3D printing. This is because, having been around for such a long time, STL is compatible with most 3D CAD software and other software and hardware in 3D printing. The STL is made from triangular *face elements (facets)* and it can be ASCII (textual) or binary format. *Triangulation tolerance, adjacency tolerance, and angle control* are some of the three important factors that control the accuracy of this file format. STL does not carry *unit information*. It also does not contain *free-form curves, color, and texture information.*

The STL file format (Figs. 4.22 and 4.23) became a standard format after being developed in the 1990s for SLA equipment. The STL file format removes any of the extra data normally left in a design by a CAD program while converting polygons into simpler triangular shapes. The size of the triangles helps determine the accuracy of the build. For example, the smaller the triangles are, the more the actual build will exactly correspond with the product design. However, most AM machines also have a maximum resolution at which they are capable of operating, which will also affect a product's quality.

Most CAD programs can convert CAD designs to STL. However, errors may occur during file conversion. Various programs can scan the STL file for any discontinuities in the design or for geometric data that may be too complex for the AM machine. Using these programs can help guarantee that the product's build is successful. Some new additive manufacturing file formats, including *AMF (Additive Manufacturing File Format)* and *3MF (3D Manufacturing Format)*, have recently been developed which vastly improve the somewhat antiquated STL format, as they add more information to the file, including color and material, and allows the use of curved triangles to improve model quality [12].

Like STL, AMF stores geometry data using triangular tessellation. However, the triangles in AMF can curve, resulting in accurate data representation. This also results in much smaller file sizes as a smaller number of triangles are required to accurately represent curved surfaces. The AM file format (AMF) standard is jointly managed by ASTM

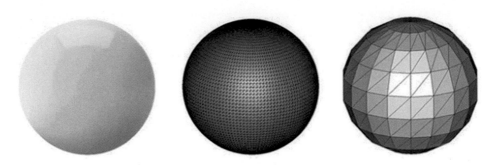

Fig. 4.23 The STL describes surface geometry of a three-dimensional object with triangular tessellations

(American Society for Testing and Materials) and ISO (International Organization for Standardization) [5].

In Practice

Draw an iron angle shaped part (with dimensions of $4 \times 4 \times 1.5$ in.) shown in the figure in a CAD program of your choice. It also should have a length of 1.5 in. Please add the through hole shown in the figure. It has a diameter of 0.75 in. and is located at the center of the face.

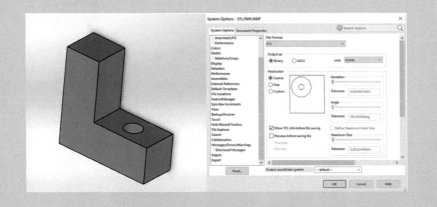

By following the options under File>Save As>.stl sequence:

- Use the "Coarse" settings with both Binary and ASCII options to save the .STL file. Record the number of triangles and file size.
- Use the "Fine" settings with both Binary and ASCII options to save the .STL file. Record the number of triangles and file size.

Gather your results in a table including the deviation tolerance, angle tolerance, and associated triangle count and file size.

3MF, or 3D Manufacturing Format, is a file format developed and published by the 3MF Consortium (Fig. 4.24). It is an *XML (Extensible Markup Language)* data format designed specifically for AM, and it includes information about materials, colors, and other information such as texture, material, and orientation data, that cannot be represented in the STL format. 3MF allows for the designer to provide all of the information and design intent that is necessary to produce a 3D printed part and is highly accurate. 3MF natively enables efficient arrangement of parts on a build platform while maintaining workflow flexibility. 3MF has all the technical properties of AMF. It also uses curved triangular tessellations to encode geometry.

Using the 3MF file format, the printer operator can map actual materials and support structures so the part will print correctly on the specific printer in use. 3MF production extension enables manufacturing traceability by uniquely identifying all objects in the build platform. 3MF's unique extensibility mechanisms allow enriching the data with custom manufacturing information. This enables integration of AM data into existing manufacturing workflows.

In general, 3MF is a much more modern format than STLs or VRMLs. It provides a smaller file size than an equivalent resolution STL (since 3MFs are zipped internally) (Figs. 4.25 and 4.26). It also provides all the texture images internally, so you can't lose them. Like AMF, 3MF appears to be gaining considerably more traction. It is adopted by

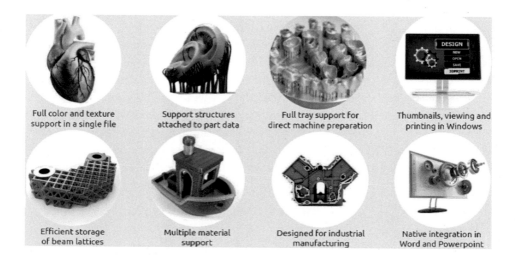

Full color and texture support in a single file	Support structures attached to part data	Full tray support for direct machine preparation	Thumbnails, viewing and printing in Windows
Efficient storage of beam lattices	Multiple material support	Designed for industrial manufacturing	Native integration in Word and Powerpoint

Fig. 4.24 3MF advantages [13]

	3MF	STL	OBJ	VRML
Always print-ready	✓	⊗	⊗	⊗
Unit aware	✓	⊗	⊗	✓
Full color capability	✓	⊗	✓	✓
Textures in one file	✓	⊗	⊗	⊗
Tray support	✓	⊗	⊗	✓
Contains support structures	✓	⊗	⊗	⊗
Unicode aware	✓	⊗	⊗	✓

3MF is tailored for Additive Manufacturing. It addresses the challenging details of Additive Manufacturing workflows.

Based on open packaging conventions, its file size offers a compact representation of your 3D Printing Data. The following example consists of 320.000 triangles and includes support geometry.

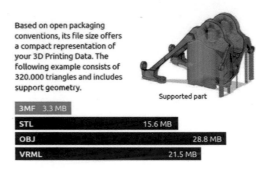

Supported part

3MF	3.3 MB
STL	15.6 MB
OBJ	28.8 MB
VRML	21.5 MB

Fig. 4.25 3MF Feature (left) and File size (right) comparison [13]

33 companies including Autodesk (Autodesk Fusion 360, Meshmixer, and Netfabb), Dassault Systems (CATIA/3D-Experience and SolidWorks), Microsoft (Microsoft 3D Builder, Paint 3D, and Office 365), SLM Solutions, HP (HP Jet Fusion 3D Printing Solution), Shapeways, Materialise (Materialise 3-matic and Magics), 3D Systems (3DXpert and 3D Sprint), Siemens PLM (Solid Edge and NX), PTC Creo, Rhinoceros, MasterCAM, Simplify 3D, Ultimaker Cura, nTopology and Stratasys [13, 13].

OBJ file format was developed by Wavefront Technologies for its Advanced Visualizer animation package. With the development of *multicolor and multi-material printing*, the file format was later adopted by the 3D printing industry. In this file format, fine triangular mesh approximately encodes the surface of the 3D geometry. Free-form curves/surfaces are included as well as color and texture information. It allows to create tessellations using various shapes such as polygons and quadrilaterals, and not just triangles to improve accuracy. In terms of popularity, OBJ is second only to STL. However, unlike STL which only stores geometry data, OBJ can store geometry, color, texture, and material data. Color data is stored in *a separate companion MTL (Material Template).* An OBJ file has to be shared with its corresponding MTL file for color printing to be possible [16].

It is generally recommended to use STL file formats for simple geometry and single-color prototypes. If you intend to print simple parts in color, then the OBJ is a better option. The 3MF and AMF 3D printing file formats are the most technically superior as they both store every information on a model. They are excellent for complex multi-part, multicolor, and multi-material objects. They also standout for their ease of file sharing as all data and metadata are stored in compact, compressed files [16].

In Practice
Open an ASCII.stl file and document how it defines the part geometry. Open an OBJ file, document its structure and elements in detail.

Fig. 4.26 To print logos and textures on simple faces of CAD models, exporting a 3MF from SOLIDWORKS is one of the simplest and most modern ways to get the information into GrabCAD Print [15]

4.5 Slicing and Support Generation

The next phase after 3D CAD model preparation is digitally slicing the 3D model into hundreds or thousands of horizontal layers, this is performed with "slicer software" prior to printing, no matter what 3D printing technique is used, this step would be necessary and important. Slicer software will also handle the "*fill*" of the model by creating a lattice structure inside a solid model for extra stability if required (Fig. 4.27). The slicer software will also add in support columns, where needed. These are required because the part cannot be laid down in thin air, and the support columns help the printer to bridge the gaps. These columns are then later removed during the finishing process. Support structures may be a vital component in certain types of 3D printing. They are usually needed for areas like overhangs, holes, and angles. However, it is always preferred to have as minimum support as possible. If the part has a certain angle that becomes close to the horizontal, you're going to need to build a secondary structure to support that overhanging feature. And the criteria for support depend on the geometry, the orientation of the part, the material, and even the 3D printing machine. The slicing layers in an FDM model are determined by the extrusion-die diameter, which typically ranges from 0.05 to 0.12 mm (0.002–0.005 in.). This thickness represents the best achievable tolerance in the vertical direction. In the X–Y plane, dimensional accuracy can be as fine as 0.025 mm (0.001 in.) if a filament can be extruded into the features. It is also to be noted that the angular support structure for horizontal holes is preferred. By using this, one can minimize the contact area with the support resulting in less post-processing. In some instances, light tubular support is preferred which takes less time to get rid of.

The cured resin is denser than the liquid resin in the Stereolithography apparatus (SLA) 3D printing process, so a large support structure is often needed to support the part. Since SLA requires a single resin material in the bath, support materials on the part must be manually removed instead of chemically removed. Therefore, the part orientation must minimize the surface area of attachment between the part and the support. The software for SLA printers, including the Formlabs Form 2 machine, often requires that the best orientation for the build is at an angle.

Selective Laser Sintering (SLS) 3D printers fuse powdered material in a chamber using a laser. For SLS, there's no need for support structures since the powder acts as a form of support when the object is built up layer by layer (Fig. 4.27). This gives a lot of design freedom but also generally increases the cost and time to print a part. Similar to SLS, there's no need for support structures with binder jetting since the powder supports the object as it's built.

Metal 3D printing technologies (SLM, DMLS) use support structures to keep models fixed to a base plate during the building process. However, overhangs with an angle greater than 35° can be printed without support. When you do need support for metal 3D printing, it's important to ensure that they are easy to access, or else it'll be challenging and maybe even infeasible to remove them during post-processing. Using supports won't

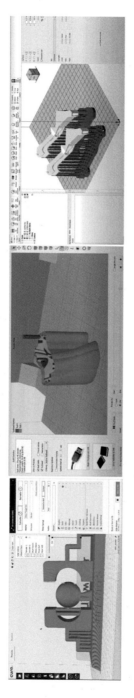

Fig. 4.27 Commonly available slicer software (Cura (left), Simplify3D (center), NetFabb (right) [17]

impact the overall quality of your part, and with the proper post-processing methods, you can remove all marks from the printed model (Fig. 4.28 and Table 4.2).

Material Jetting (Stratasys PolyJet and 3D Systems MultiJet Modeling) technologies are similar to inkjet printing, but instead of jetting drops of ink onto paper, these 3D printers jet layers of liquid photopolymer onto a build tray and cure them instantly using UV light. These printers require the use of support material in all cases where there are overhanging features, regardless of the angle. Supports are either water-soluble or are removed during post-processing using pliers, water jetting, ultrasonic baths, or sandblasting.

Fig. 4.28 SLS prints without support structure

Table 4.2 Summary of whether support is required for each of the 3D printing technologies

3D printing technology	Need support structures?
FDM (desktop and industrial)	Depends on model geometry
SLA and DLP	Always
Material jetting	Always (dissolvable)
SLS and MJF	Never
Binder jetting	Never
Metal 3D printing	Always

Support structures will generally affect the cosmetic appearance of a part, therefore post-processing is needed to improve the surface finish after removing supports. Material Jetting is the exception to this rule. The more support structures you print, the more complex a design can be for certain 3D printing technologies. The amount of support material used can be optimized by addressing part orientation and part accuracy (among other design and manufacturing factors) to lower the cost and print time.

So, any reasonably good 3D slicer software will create:

1. A toolpath based on the geometry of your STL file.
2. A percentage of infill to save 3D printing time and material.
3. Generation of support material if the geometry is challenging to print. These supports are meant to be removed after the print is finished.

Once a STL file has been generated, the file is imported into a slicer program, which slices the design into the layers that will be used to build up the part. The software will slice up the part according to the layer increments specified in the modeler setup. After the file sliced, the next step is to generate toolpaths.

Sample slice files in proprietary Catalyst and Insight (Courtesy of Stratasys) given in Fig. 4.29 [7]. The slice file is like a stack of transparencies. Each transparency has the interior and exterior boundaries of that horizontal cross section. With toolpaths generated, the next step is to identify and fix any problems in the slices. Not all.STL files are clean and error-free. Problems can arise from a couple of different sources. First, some CAD systems are better at creating STL files than others, so they are commonly used and the problem may have been caused by them. Secondly, the designer may not have been using good design practices. If the designer did not create watertight (i.e. manifold) surfaces, the problems will appear in the sliced file.

These problems can be eliminated by software packages like Meshmixer, Netfabb, MeshLab, Cura, PrusaSlicer, Simplify3D, etc. [10, 10]. Because the bunny model given in Fig. 4.30 is a 3D model that was created by 3D scanning a real-world object, the scanner was only able to capture the exposed surface areas of the bunny, and not the parts where it was sitting on the table. To 3D print this bunny, this "missing and open" part of the scan will need to be closed. 3D printing requires a closed, airtight model. Meshmixer has a powerful *Inspector* tool that will evaluate and "fix" 3D models to make them more 3D printable. Sometimes more advanced repair is needed, but often the Inspector tool does a great job automatically.

After analyzing the file and offering choices and settings, the slicer software generates a "G-code" file that's tailored for the specific 3D printing machine for tool paths. It describes coordinates, nozzle and bed temperatures, fan control, printhead speed, and other variables. After slicing, the data is sent to the printer for the final stage via USB, SD, or via the Internet. Some 3D printers have a built-in slicer and allow user to feed supported file formats like STL, OBJ, AMF, 3MF, etc. which they slice themselves. The

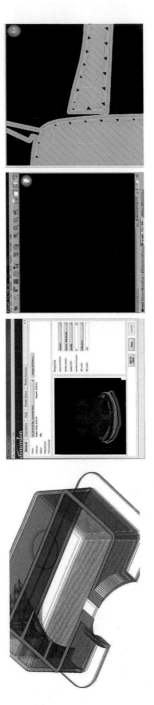

Fig. 4.29 An example slice file in Catalyst and Insight (Courtesy of Stratasys). Toolpaths generated (rightmost- green interior) [7]

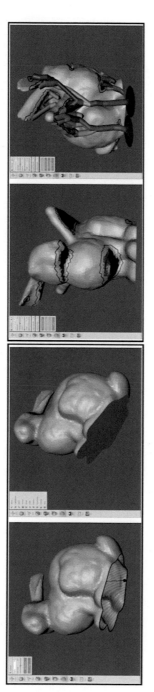

Fig. 4.30 A blue "pin" popping out of the bottom of the bunny (left picture) indicates that there is an issue (hole) with the mesh in that location. Meshmixer will find errors such as holes and small disconnected parts with the Inspector tool. (left) Meshmixer can analyze overhangs and create support structures to keep 3D model in place during the printing process [10, 18]

slicer program takes the and converts STL file into G-code. G-code is an NC programming language used in CAM (Chap. 3) to control automated machines like CNC machines and 3D printers [19, 20].

The slicer program also allows the 3D printer operator to define the 3D printer build parameters by specifying support location, layer height, and part orientation (Fig. 4.27). Slicer programs are often proprietary to each brand of 3D printer, although there are some universal slicer programs like Netfabb, Simplify3D, Slic3r (Fig. 4.27), and Ultimaker's CURA. Along with the above information, G Codes or similar tool (laser) positioning instructions, recoating instructions (powder recoating) etc. are included in the file, which the 3D Printer then follows and performs the actual process which could be lasers curing resins or extruding plastic through a hot nozzle etc.

G-Codes for AM

G-code is a programming language for computer numerical control (CNC). In other words, it's the language spoken by a computer controlling a machine, and it communicates all commands required for movement and other actions. G-code became available with the emergence of open-source machines, and especially FFF 3D printing and includes sequential lines of instructions, each telling the 3D printer to perform a specific task. These lines are known as commands, and the printer executes them one by one until reaching the end of the code. While the term "G-code" is used to reference the programming language, it's also one of two types of commands used in 3D printing: "*general (preparatory)*" and "*miscellaneous*" commands. General command lines are responsible for types of motion in a 3D printer. Such commands are identified by the letter 'G', as in G-commands. Besides controlling the three plus axes movement performed by the printhead, they're also in charge of filament extrusion. The miscellaneous commands, on the other hand, instruct the machine to perform non-geometric tasks. In 3D printing, such tasks include heating commands for the nozzle and bed and fan control, among many others, as we'll see. Miscellaneous commands are identified with the letter 'M' (Fig. 4.31).

Review Questions

1. Define the terms "rapid prototyping (RP)", "rapid tooling (RT)", and "additive manufacturing (AM)".
2. List the 9 steps of the AM workflow.
3. What are the two ways of achieving a design? (Hint: Remember reverse engineering)
4. What is Design for Additive Manufacturing (DfAM)?
5. What are the major components of DfAM? Explain each with an example.
6. List 5 design constraints of AM. Explain each one concisely.
7. Compare and contrast the four major intermediate file formats used in 3D printing including. stl,.obj,. amf, and 0.3mf.
8. How can moving assemblies be handled in pre-processing including rotating ones?

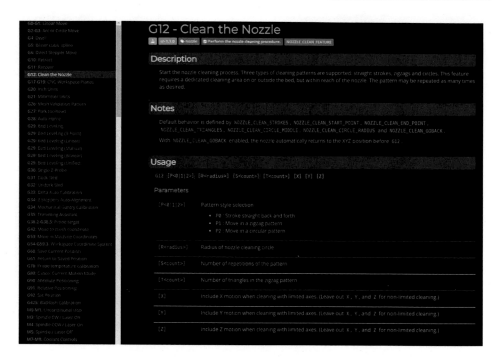

Fig. 4.31 Typical G and M codes used in 3D printing [21] (See also Appendix II)

9. Why do FDM-made parts exhibit anisotropy: strong direction(s) and a weak direction(s) where when stress is applied the part fails?
10. How is an .stl file sliced? Explain concisely.
11. Which 3D printing method never requires support structures?
12. Which 3D printing method will always require support structures?
13. What is the difference between infill patterns and lattice structures?
14. What is topology optimization? By using the example given in the chapter, explain it.

Research Questions/Challenges

1. Consider this piece below for questions 1–3. Why did the designer choose to orient the part in this way? Which considerations did the designer likely keep in mind?

2. Which of the following actions can help to decrease the build time of a part?
   Select all that apply. There is more than one correct answer.
   a. Minimizing support structures
   b. Intertwining small features with support material
   c. Maintaining air flow across the part
   d. Maximizing horizontal area to minimize Z
3. As material leaves the extrusion tip, it should be...
   a. Liquid, so that it can bond with other layers
   b. Just under solidification point
   c. Just above solidification point
   d. Fully solid, or it will collapse
4. Which of the following support styles are used for soluble supports?
   Select all that apply. There is more than one correct answer.
   a. Smart
   b. Sparse
   c. Basic
   d. Breakaway
5. Place these steps for creating a 3D printer file in the correct order.
   a. Orient the part
   b. Generate supports
   c. Send to Control Center
   d. Open the.stl File
   e. Generate toolpaths
   f. Fix errors and problems
   g. Slice the file
   h. Configure the modeler setup

    i. Create Toolpaths
6. Compare the two model orientations below for surface finish and strength characteristics.

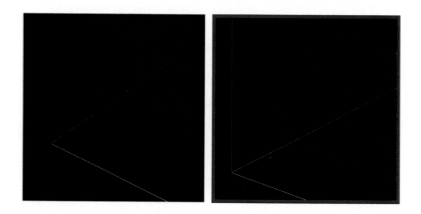

7. Explain how the user controls the tolerance of a triangulated model (.stl) by inputting important data. (Hint: Consider parameters like chord height, triangulation and angle tolerance etc.)
8. Explore Ultimaker's CURA software's surface mode and its non-manifold print ability. Try to print a non-manifold model of your choice.

Discussion Questions

1. The economics of machining and 3D-printing are different. For 3-D printing, some have said "complexity is free." Explain concisely.
2. How would you improve below design for AM? Show your improvement with a sketch?

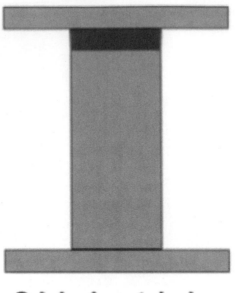

Original part design

3. Consider the CAD drawing below, and select the design adjustments needed to make this part free of support while printing.

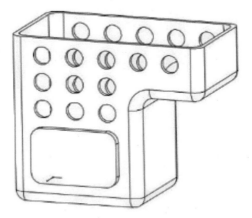

4. What design improvements are made possible in the parts presented here?

5. How would you change the design given below to save model material and printing time?

References

1. Hopkinson, N., Hague, R.J.M., Dickens, P.M. (2006). *Rapid Manufacturing: An Industrial Revolution for the Digital Age*, John Wiley & Sons.
2. Chua, C.K., Leong, K.F. and Lim, C.S. (2003). *Rapid Prototyping, Principles and Applications*, World Scientific Publishing Co. Pte. Ltd, Singapore.
3. Liou, F. (2019). *Rapid Prototyping and Engineering Applications 2nd Edition*, CRC Press.
4. Harvard Business Review: The 3D Printing Revolution, https://hbr.org/2015/05/the-3-d-printing-revolution, accessed September 1, 2022.
5. ASTM: The 5 Most Important Standards in Additive Manufacturing, located at: https://www.astm.org/standardization-news/?q=features/5-most-important-standards-additive-manufacturing-.html/accessed September 1, 2022.
6. Martin Leary, (2019), *Design for Additive Manufacturing*, Elsevier Science.
7. Stratasys, located at: www.stratasys.com/, accessed September 1, 2022.
8. Hubs, located at: https://www.hubs.com/, accessed September 1, 2022.

9. DeGrosky, J., Workmaster, C., Dodds, N., McChesney, C., Minarik, N., Proctor, B., Leimkuehler-Mullin, L., Leimkuehler, P., Sirinterlikci, A., Carlsen, R., Joo, W. (2022). Adopting Additive Manufacturing Technologies for Orthotics and Prosthetics, Unpublished Robert Morris University Research and Grants Expo Poster.

10. Meshmixer, located at: https://www.meshmixer.com/, accessed September 1, 2022.

11. Hull, Chuck "On Stereolithography" (2012), Virtual and Physical Prototyping. 7 (3), pg.177.

12. Lipson, H., Additive manufacturing file formats, located at: https://fab.cba.mit.edu/classes/S62.12/docs/formats.ppt, accessed November 23, 2018.

13. 3mf datasheet 072020, located at: https://3mf.io/, accessed September 1, 2022.

14. 3mf adoption, https://3mf.io/3mf-adoption/, accessed September 1, 2022.

15. How to export your color Solidworks files for printing, located at: https://grabcad.com/tutorials/how-to-really-export-your-color-solidworks-files-for-printing, accessed September 1, 2022.

16. 3D printing file formats compared obj, stl, amf, and 3mf, located at: https://xometry.eu/en/3d-printing-file-formats-compared-obj-stl-amf-and-3mf/#:~:text=In%20terms%20of%20popularity%2C%20OBJ,companion%20MTL%20(Material%20Template), accessed September 1, 2022.

17. Best 3D slicer software 3D printer, located at: https://all3dp.com/1/best-3d-slicer-software-3d-printer/, accessed September 1, 2022.

18. STL repair fixer tool online offline, located at: https://all3dp.com/2/stl-repair-fixer-tool-online-offline/, accessed September 1, 2022.

19. What is 3D printing, located at: https://3dprinting.com/what-is-3d-printing/, accessed September 1, 2022.

20. How exactly does 3D printing work, located at: https://interestingengineering.com/how-exactly-does-3d-printing-work, accessed September 1, 2022.

21. G012, located at: https://marlinfw.org/docs/gcode/G012.html, accessed September 1, 2022.

3D Printing Processes and Associated Materials

5

3D Printing processes of today are mainly additive in nature where the raw material is processed, deposited, or cured in layers. Even though a 3D object is generated through printing, most operations only control 2 of the machine's axes (x and y) concurrently before the 3rd axis (z) is adjusted for the next layer. On the contrary, the *non-planar (or true)* 3D printing processes allow simultaneous *control of all of the three axes*, producing curvature that stretches across them. Fused Filament Fabrication (FFF) version of this type of printing is relatively new, and stemmed from the work done at the University of Hamburg in Germany in 2018 [1]. Processes combining additive (material deposition) and subtractive (material removal) have been in existence as well, even though they have not gained popularity including the *3 in 1 machines with milling, laser engraving, and 3D printing ability.* Figure 5.1a shows a DENFORD Compact 1000 router with a manually attached FFF 3D printhead replacing its spindle column. This project was completed in 2013 at RMU (Fig. 5.1b). A laser engraver attachment was later designed, making this machine a 3 in 1 [2].

3D Printing processes can be classified in a variety of ways including their:

- Energy source, i.e. electron beam, laser, or light source
- Nature of the process, i.e. binding powder materials with chemicals, curing of liquid materials, melting or sintering of powder materials. There are 7 categories of additive manufacturing (or 3D printing) methods based on the *American Society of Testing and Materials (ASTM) F42 committee*'s work that classify the range of the processes into categories [3]:
  - *VAT polymerization:* is based on exposing liquid photopolymers in containers (vats) to UV light or laser to convert them into solids by a chemical reaction.

© The Author(s), under exclusive license to Springer Nature Switzerland AG 2023   133
A. Sirinterlikci and Y. Ertekin, *A Comprehensive Approach to Digital Manufacturing,*
Synthesis Lectures on Mechanical Engineering,
https://doi.org/10.1007/978-3-031-25354-6_5

(a)

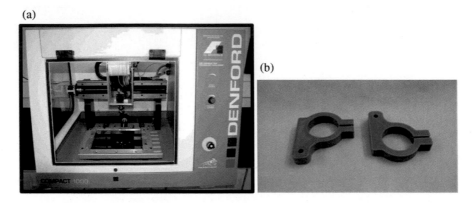

(b)

Fig. 5.1   **a** A DENFORD router modified to be a 3 in 1 machine **b** Prints from the machine [2]

- – *Material jetting:* where liquid photopolymer material is jetted onto a build platform and cured by being exposed to UV light. Jetting process can be done continuously or be based-on *drop-on-demand (DOD)* technology.
- – *Binder jetting:* is a process in which the printhead selectively deposits a liquid binder onto thin layers of powder particles.
- – *Material extrusion:* is based on heating and extruding a continuous thermoplastic filament onto a build plate to generate the part in layers.
- – *Powder bed fusion:* Employs a heat source including a laser or electron beam to sinter or melt powder particles in layers.
- – *Sheet lamination:* is a process where thin sheets of material are fed by rollers and bonded together layer-by-layer to form a single piece that is cut into a 3D form.
- – *Direct energy deposition:* A method that deposits and melts a material in wire or powder form onto an object surface by using a laser or electron beam, or plasma arc.
- • Raw material form, i.e. liquid, solid, paste, plasma, powder, spray
- • Build volume size, i.e. large, medium, small, or micro
- • Purpose, i.e. industrial, benchtop/office, or hobbyist.

This book uses the classification based on the form of the raw material: solid-, liquid-, and powder-based systems. A fourth category, called other, includes material forms like paste, plasma, or spray.

5.1   Solid Raw Material Based Processes

Solid raw materials are employed in a variety of 3D printing technologies, mainly in the form of a filament, sheet, or rod in a cartridge. Besides material extrusion, early 3D printing technologies included fabrication of 3D metal parts by lamination. The layers of these parts were CNC machined and then *laminated* via *brazing* at the Ford Motor Company. In a similar fashion Nakagawa [4] laser cut the layers of his sheet metal tooling and joined them with *diffusion bonding*. His method has especially gained some popularity over the years [5]. The following section covers the processes with solid raw materials including:

- Fused Deposition Modeling (FDM)/Fused Filament Fabrication (FFF)
- Continuous Filament Fabrication (CFF)
- Bound Metal Deposition (BMD)
- Wire-Arc Additive Manufacturing (WAAM)
- Ultrasonic Additive Manufacturing (UAM)
- Laminated Object Manufacturing (LOM).

5.1.1   Fused Deposition Modeling/Fused Filament Fabrication

Fused deposition modeling (FDM) process is originally based on heating a thermoplastic polymer filament to beyond its glass transition temperature where the *material changes its state from hard and glassy (state of amorphous materials) into soft and malleable, before it is extruded in layers as shown* in Fig. 5.2. The process was invented and patented by S. Scott Crump, the co-founder of Stratasys, in 1992 [6]. The lead author was introduced to the FDM 1600 machine in 1994 at the Ohio State University. At the time, the machine cost about $100 K, ran on Unix workstations, and was able to print with acrylonitrile–butadiene–styrene (ABS) and wax materials. While the lead author was engaged in printing prototypes (i.e. a whistle) for local industries, he also worked on a Ford Motor Company project with an aim of printing spare parts for old car models in place of following the rather costly route of molding. This very early additive manufacturing effort failed in late 1994 due to low quality of the prints, including their surface quality.

The FDM process requires support structures to be printed along with the part being printed to assure a successful outcome. Three support types are illustrated in Fig. 5.3a and they are:

- To separate the prints from the build plate to assure easy and safe removal of them as well as preventing warpage due to part pulling away from the build plate locally
- To prevent holes from shrinking.
- To hold overhangs.

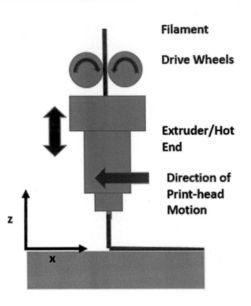

Fig. 5.2 The FDM process

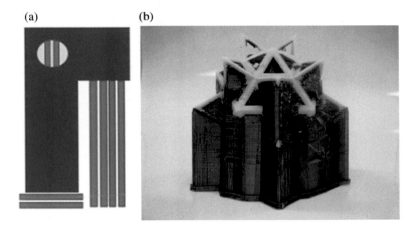

Fig. 5.3 **a** Support structures [8] **b** A Stratasys Dimension Elite FDM print with water-soluble supports

- Overhangs that are making 45 degrees or more (with the vertical axis) will need supports (i.e. 90 degrees in Fig. 5.3a) [7].

Over the years, the FDM process saw development of *water-soluble* (i.e. Ultimaker's polyvinyl alcohol (PVA)) or *biodegradable* (Stratasys ABS-like) *supports (Fig. 5.3b)* and *composites*, and with the Rep-Rap movement [9], first the open-source machines running on G-codes (i.e. Mendelmax 1.5) came along and then the low-cost machines that cost only a couple of hundred dollars (i.e. Creality Ender 3). Programs like Slic3r (used in slicing the .STL geometry and generating the.gcode) and Pronterface (3D printer interface) became household names. After the expiration of the FDM patent in 2006, the process became available to many, kits transforming into low-cost machines over time. However, these machines cannot be called FDM machines due to the Stratasys' trademark on them, they are labeled with the term, *fused filament fabrication (FFF)*. The Ultimaker's Cura software is now common for FFF printers after replacing its competitors since it allows working with any FFF printer [7]. Figure 5.4a, b show two different FFF machines, one with an enclosure and the other without an enclosure. Early Rep-Rap and FFF machines did not have enclosures, but the need of controlling the temperature within the build volume and achieving printing of higher temperature materials making enclosures a necessity and more common (Fig. 5.5a).

With the emergency of FFF printers, the role of support structures has also been expanded to include skirt and brim, and the interface between the build plate and part was also called with a new term, raft (Fig. 5.6) [7].

(a)   (b)

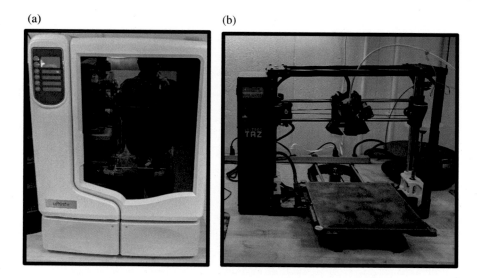

Fig. 5.4 **a** A stratasys uPrint SE FDM machine **b** A LulzBot TAZ 6 FFF machine

(a)   (b)

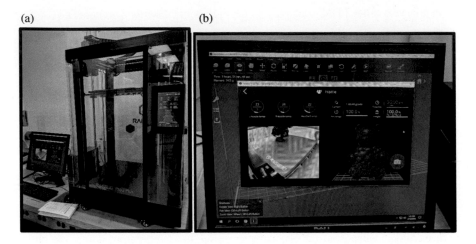

Fig. 5.5   a Raise Pro 2 3D. **b** Raise's print monitoring also includes time-lapse videos of the successful prints

(a)   (b)   (c)

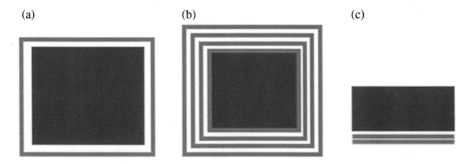

Fig. 5.6   a Skirt, **b** Brim, and **c** Raft (raft varies along the z-axis even though this was not depicted in the figure [8])

- *Skirt* is the default option in Cura (Fig. 5.6a). It is a line around the print on the first layer that helps to prime the extruder. If the print shows significant warping, the user should consider changing to the other options, Brim or Raft.
- When the user chooses *brim*, Cura places a single-layer-thick and flat area around the print (Fig. 5.6b). This layer resists the pulling forces as the print cools, stopping the warpage. Since the brim is made from only a single layer thick, it's easy to remove once the print is finished.
- In some instances, brim might not be enough to prevent warping, using a raft becomes necessary (Fig. 5.6c). A raft adds a thick support grid between the part and the build

plate, ensuring that the heat is distributed equally as it makes removal of the part better as explained earlier. Raft is particularly critical when the bottom of a model is not completely flat, or when printing with industrial grade materials.

Additional precautions can also be taken in setting printing process parameters to achieve good prints.

The most commonly used materials today in this process include semi-crystalline polylactic acid (PLA), amorphous glycol-modified poly (ethylene terephthalate) (PETG), amorphous acrylonitrile–butadiene–styrene (ABS), semi-crystalline Nylon 12 (PA 12) and others such as amorphous polystyrene (PS), amorphous polycarbonate (PC), amorphous ULTEM® (polyetherimide). Medical grade and sterilizable materials are also available. Often these materials are prone to collecting moisture from their environment, thus *dry-boxes* have become a critical part of the FDM/FFF processes. In one case, while working on printing filter end-caps for a hydraulic fluid conditioning company, Schroeder Industries located in Leetsdale, Pennsylvania, the lead author observed steam coming out of an FFF extruder as the print was progressing due the filament's moisture absorption.

The lead author has been involved in development of various FFF composite materials including polymer-metal composites delivering antimicrobial effect since 2012 [10]. This is another trend seen in the FDM/FFF realm. These composites can be used to fabricate polymers with electrical conductivity or elevated strength. Wood polymer (PLA) composites have been on the market too, originally envisioned for architectural 3D printing. They include (i.e. LayWood-Flex filament) wood fibers or flour (up to 35%), a polymer binder, and a flexibilizer [11]. FFF/FDM Composite material printing has expanded beyond the research and hobbyists' spaces. Markforged's Onyx machine was released in 2016. It's newer version (Onyx One™) today also prints in 100–200-micron layers with Onyx material, a combination of nylon and chopped carbon fibers. This material with a tensile strength of 37 MPa is 1.25 times stronger than ABS prints done with the same process in addition to being 1.1 times stiffer with a Young Modulus of 2.4 GPa [12]. This type of reinforced materials has also low density, and they are resistant to corrosion, static electricity, and high temperature. Raise3D also recently introduced an FFF printer, E2CF with a carbon fiber reinforced Nylon material (PA12CF). Besides printing a composite material, E2CF has two independent print heads (*an IDEX* printer) and only costs around $4.5 K [13].

In Practice
A part was designed to study the "Convex Hull" method in determining 3D printing build orientation based on a study conducted by Zwier and Witts [14]. Meshlab software was used in the generation of convex hull of the object, wrapping around it as shown below with its red edge lines. Any of the design's surfaces coinciding with the convex hull are the potential surfaces of placement for 3D printing. After

completing the review of this case, please duplicate this study with a design of your choice by employing the software tools utilized.

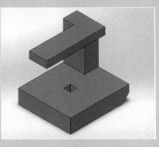

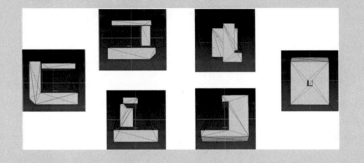

Following 3 orientations were obtained from the convex hull method where each placement surface coincided with the convex hull wrapping around the model. As expected, they resulted in the 3 best results of all possible orientations in terms of building time and total material used including the amount of supports:

1. 66.94 g, 6 h 45 min
2. 70.81 g, 7 h
3. 91.19 g, 8 h 49 min.

In Practice

Following the instructions given below for Ultimaker's Cura, prepare a print for an FFF 3D printer.

- Open the .STL file you want to print using the "Open File" button (the first button located in the top left corner) from the "Stage Menu", marked in the figure as the "Prepare" tab.

- Once the model is shown in the 3D viewer of the software, user can relocate the model by rotating and panning (translating). User can also scale, and duplicate the geometry on the build plate.

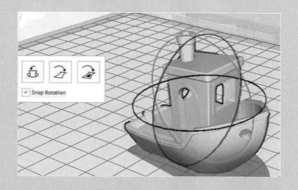

- Then, use the second button ("Printer Selection Panel") in the "Stage Menu" to select the 3D printer of your choice i.e. (Ultimaker) S3 or (Creality) Ender 3. Network enabled printers exhibit a valid connection with a checkmark icon. (You can also transfer the.gcode to a non-networked printer with an SD card.)
- The third button in the "Stage Menu" is the "Configuration Panel" including the current material setup/print nozzle for the selected printer i.e. Black PLA or Natural PVA and 0.4 mm. While networked printers will exhibit their available configurations, non-network printers need to be configured manually.

- The rightmost button in the "Stage Menu" includes the "Print Settings Panel" for all settings required for the printing process. This panel opens in the recommended defaults allowing user to do a quick print with optimized printing profiles. The panel is comprised of multiple settings/parameters:

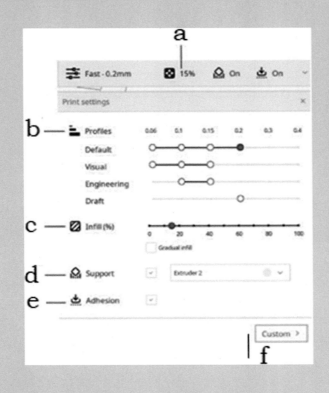

a. "Print settings panel": Demonstrates the current print settings.
b. "Print profiles": The available profiles (with the layer height slider) for the current configuration are marked with a blue dot (i.e. 0.2 mm).
c. "Infill slider": Helps adjust the model density and its strength i.e. 15% marked with a blue dot.
d. "Support selection": Enables/disables automatically generated support structure assigning them to an extruder i.e. Extruder 2 on Ultimaker S3.
e. "Adhesion": Enables/disables additional adhesive printing as the type of the adhesion is automatically set by the print profile.
f. "Custom mode": Allows custom setting of the parameters.
   - Once the printer type, configuration and print settings are in place, the model can be sliced using the "Slice" button in the bottom right corner of

the user interface. This move the process into the "Preview" stage. Please try the "Layer Slider and Simulation View" to inspect the model's print in layers.

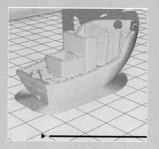

— You can then print via cloud using "Ultimaker Digital Factory", save to an SD card or save the files to the Digital Library.

5.1.2 Continuous Filament Fabrication

Markforged's Onyx Pro machine is the basis for the continuous filament fabrication (CFF) process [15]. It was released in 2017 after the release of its predecessor in 2016. The original CFF process employed a nylon filament with chopped carbon fibers, to be printed as the layers (100–200 μ in thickness) of the part utilizing the first of the two print heads—Fig. 5.7 (shown on the right) and Fig. 5.8a. The second print head (Fig. 5.7, shown on the left and Fig. 5.8b) introduced continuous strands of the fiberglass or possibly other reinforcing materials (Kevlar or carbon fiber) in between the layers. The parts are printed with a tessellation pattern as the infill, optimizing the strength, weight, and the print quality

Fig. 5.7 Markforged Onyx
Pro print heads

(Fig. 5.8c). In addition, the continuous fiber replaces the need for the conventional infill in some geometries improving the performance of the prints, including their strength. Thus, the system can substitute the normal infill of a part with automatically routed continuous fibers. The strength increase of the part depends on four factors: the print orientation, the amount and location of the added continuous fibers, as well as the fiber type. The following Table 5.1 summarizes the available fiber types and their uses [16].

Besides the four available fiber types, there are three plastics currently available including Onyx, Precise PLA, and Smooth TPU. According to the company, the prints with fiberglass yield a tensile strength of 590 MPa (19.0 times that of ABS, 1.9 times that of 6061-T6 Aluminum) and has a Young's Modulus of 21 GPa (9.4 times that of ABS, 30% of that of 6061-T6 Aluminum) [15].

5.1.3 Bound Metal Deposition

Bound Metal Deposition™ (BMD) is developed by Desktop Metal (DM). It is an extrusion-based process similar to FDM/FFF with one difference. Its printing material is made from powder-filled rods (Fig. 5.9a) being presented to the print head from cartridges. Desktop Metal refers to these rods as powder-filled thermoplastic media, a combination of metal powder and wax and polymer binder. The rods are heated and extruded onto the build plate in layers for forming the part. The DM's Studio System™ prints parts with their supports separated by ceramic interface media, which is also in rod form (the Ceramic Release Layer™—Fig. 5.9b). Thus, during the print, the supports do not bond to the part. Once the prints are completed (Fig. 5.10a), the parts are put into the debinding equipment (Fig. 5.10b), where the most of the primary binder is removed utilizing a solvent to create an open-pore channel structure [17, 18]. Then, the sintering furnace is used to densify the print (Fig. 5.10c). Sintering will force the ceramic interface media to disintegrate with application of small force by manually separating the part and its supports including its raft. Sintered BMD parts shrinks about 17–22% during densification [17].

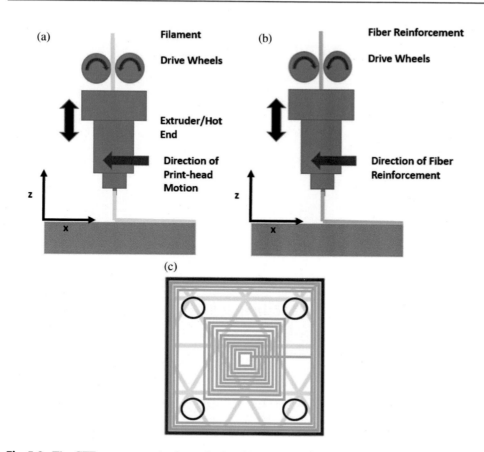

Fig. 5.8 The CFF process—**a** the first print head **b** the second print head **c** the yellow infill pattern and the green closed looped fiber enforcement

Table 5.1 CFF fiber types and their uses [16]

Fiber type	Carbon fiber	Fiberglass	HSHT fiberglass	Kevlar
Properties	High strength-to weight ratio, stiff	Sturdy, cost-effective	Sturdy, high heat deflection	Tough, impact resistant
Ideal loading type	Constant loading	Intermittent loading	Constant loading at high temperatures	Impact loading
Failure mode	Stiff until fracture	Bends until fracture	High energy absorption until fracture	Bends until deformation
Characteristics and advantages	Metal stiffness strength, lightweight	Economical starting point, general-use fiber	Keeps strength at high temperatures	High deflection and impact resistance

(a)   (b)

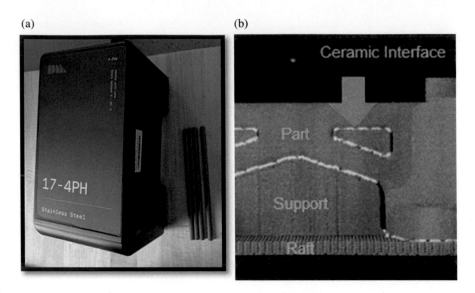

Fig. 5.9 **a** 17-4PH stainless steel cartridge for the DM's Studio System and its composite rods **b** BMD print structure

The BMD process can be applied to any sinterable powder materials that is compacted in a thermoplastic media including metallic alloys such as stainless steel, tool steel as well as metals difficult to 3D print including refractory metals, cemented carbides, and ceramics. The debinding and sintering processes need to be altered based on the chemistry of the material in concern. This process can build structures and geometries previously unattainable via bulk manufacturing processes like metal injection molding (MIM) or powder metallurgy. The process results in near-net-shape parts with the desired strength and accuracy for functional prototyping (Fig. 5.11a), tooling applications including jigs & fixtures, and possible low-volume production [17].

When compared to other metal printing processes with powder materials, the BMD process present multiple advantages:

- The Studio System does not employ wire feedstock, loose powders, and energy sources like lasers or an electron beam. It is based on extrusion of rods kept in cartridges.
- It is suitable to laboratory environments, and possibly to offices *as long as its hazards are handled appropriately including its debinding solvents as well as gasses and exhaust of the sintering furnace.*
- Support removal is much easier than other metal printing processes where machining is required due to their rigid support structures used for helping dissipate heat during processing for preventing internal stress issues.
- The process enables fabrication of parts with fully-enclosed fine voids. With the exception of extremely small geometries, all parts are printed with closed-cell infill

Fig. 5.10 a BMD Printer **b** Debinder **c** Sintering furnace—located at the RMU EIC Advanced Manufacturing Laboratory in Pittsburgh, PA

(Fig. 5.11c)—*an internal lattice structure*. Closed-cell infill is not possible with powder-based methods, such as selective laser melting (SLM), which are restricted to open-cell lattices to evacuate unprocessed powder from the voids.

- Both print and debind times are reduced by infill (internal lattice structure). The time it takes to debind a part is directly related to cross-sectional thickness, which is controlled by printing with infill. Infill also reduces the weight of a part.

Even though the BMD process does not involve melting, rapid solidification, and their negative effects, a badly designed part can still fail as seen in Fig. 5.11b as in the case of a tensile specimen. Thus, the company has published a design guide which is available on the Internet [18]. Summary of BMD guidelines include: Maximum and minimum part size, minimum wall thickness, minimum hole size, minimum pin diameter, minimum

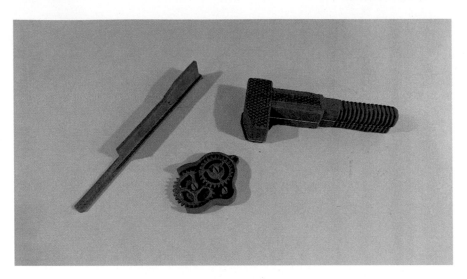

Fig. 5.11 BMD parts **a** a sintered gearbox including its casing **b** a failed tensile specimen after sintering **c** an incomplete green part exhibiting the lattice structure

embossed feature, minimum engraved picture, minimum unsupported overhang angle, minimum clearance, aspect ratio (height to width of unsupported tall thin features), layer height (50–200 microns), infill line spacing, and maximum shell thickness. Details of these will be a subject of an end-of-chapter question. Even though it was listed above within the advantages of the Studio System, the debinding process produces chemical waste and adds to the complexity of the overall process. Thus, DM has recently eliminated this step, making the process a two-step one with the new version, Studio System 2. The company also added a new isotropic *triply periodic minimal surfaces (TPMS)* infill pattern to improve the part performance. In addition to the 17-4PH stainless steel, H13 tool steel, and copper, new materials like 316L stainless steel were also to become available [19].

5.1.4   Wire Arc Additive Manufacturing

Wire Arc Additive Manufacturing (WAAM) is a direct energy deposition method based on a welding process or using an energy source like electron beam or laser in melting and depositing wire as feedstock, Fig. 5.12. It can produce large near net shape parts by depositing wire in layers to obtain the form Fig. 5.13.

Welding process in manufacturing of 3D forms dates back to 1926 when a scientist named Baker had a patent for a system that utilized electric arc welding [20]. Over the 1980s and 1990s, *shape welding process* was used to fabricate high quality large steel parts, followed by multiple patents being earned in the U.S. and Europe under the names,

Fig. 5.12 A WAAM method

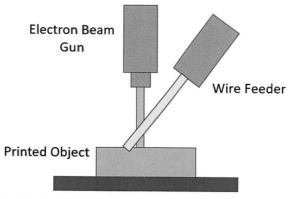

Fig. 5.13 A large WAAM part made by Gefertec GmbH (located in Berlin, Germany) being showcased in the RAPID conference

Shape Deposition Manufacturing, Shape Metal Deposition respectively [20]. Some of these efforts also incorporated CNC machining to obtain the final geometry. The process gained its popularity back over the 2010s due to the industrial needs of large AM parts. The lead author was also experimenting with a WAAM machine concept in 2014 after Michigan Technological University built one. This machine was a combination of a MIG welder with flux core and an inverted delta robot—Fig. 5.14.

The WAAM process offers multiple advantages over other similar processes as long as it has the accuracy and precision for manufacturing intent parts:

- It can build large parts since other processes (like powder bed fusion) are limited in their build volumes.
- Produces less material waste since it does not have the unused but impacted materials as well as supports.
- Offers design freedom with no requirement of having constraints like evacuating unused powders.
- May also result in reductions in lead time and costs compared to the conventional processes like casting and forging.

Fig. 5.14 RMU WAAM
machine development

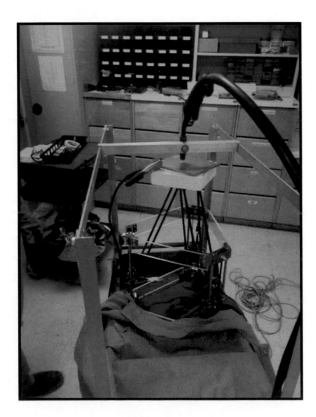

According to Renganathan, a wide range of metal alloys have been explored including titanium alloys, aluminum alloys, maraging and stainless steels, Invar ®, Inconel®, and copper alloys [20]. He also cites a few WAAM use cases including "landing gear rib, wing spar for aerospace industries, impellers, turbine blades, and shrouds and linkages for various industries" [20].

5.1.5 Laminated Object Manufacturing

Laminated object manufacturing (LOM) is a rapid prototyping system originally developed and commercialized by Helisys Inc. in 1991. Its technology used adhesive-coated paper laminates cut by a laser to make up the shape of the parts [21]. The process did not gain much popularity due to its limitations and issues including fire hazards. There was a relatively recent attempt by Cubic Technologies to bring this technology back and did not also receive commercial recognition. This attempt incorporated a knife in place of the laser and added an inkjet print head to dye the printed parts to make them look realistic. The lead author saw an original LOM machine and parts made by it during an SME event (predecessor to the RAPID conference) in 1994. The LOM parts felt just like balsa wood.

However, the lamination type of 3D printing has seen other materials being applied including plastic sheeting and sheet metal over the decades. Examples of metal lamination are given in this chapter in Sect. 5.1. Another similar effort is covered in the next Sect. 5.1.6 (Fig. 5.15).

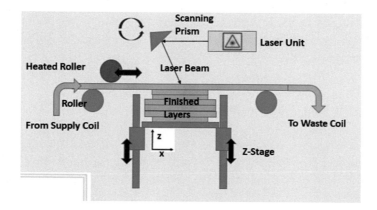

Fig. 5.15 The LOM process

5.1.6   Ultrasonic Additive Manufacturing

Ultrasonic additive manufacturing (UAM) (aka Ultrasonic consolidation (UC)) employs ultrasonic welding technology developed by Fabrisonic Inc. located in Columbus, Ohio. The company was formed in 2011 based on the Intellectual Property (IP) covering the aspects of the UAM process via nine patents. The original work was done at the Edison Welding Institute (EWI), an Ohio based research organization [22]. The process takes advantage of ultrasonic vibration, high-frequency sonic energy, to merge *featureless foil stock (tape)* into 3D forms. The process is accomplished while material is in solid-state thus, does not involve melting, solidification, and cooling. It works with an array of materials including aluminum, copper, stainless steel, titanium, and produces fully dense parts.

The process is based on the following principles and equipment [23]:

- The ultrasonic horn (with a textured surface to grip metal tape) rolls over the metal tape layer as it vibrates.
- Ultrasonic motion causes oxides between the tape layer to break down.
- Localized asperities yield and collapse.
- Heat and pressure create high strength solid state bonding.
- CNC machining is needed to complete the part into final shape.

Fabrisonic machines are made from 3 Axis CNC mills—Fig. 5.16. Fabrisonic adds its patented welding print head as another tool in the CNC's *automatic tool changer* (ATC). The print head produces a part that is near net shape. The milling features can then be used on the CNC machine to obtain the final true shape. This hybrid system yields a very smooth final surface finish as well as tolerances of $\pm 0.0005''$. This technology allows high production rates (15 to 30 in³/hour) in large build volumes, it can also be used in mending worn parts with ease.

5.1.7   Liquid Raw Material Based Processes

The history of ultraviolet (UV) curing of photopolymers started with Hideo Kodama, a Japanese researcher, who invented a layered printing technique in the 1970s [24]. In 1983, Charles Hull developed a prototyping system he called *stereolithography (SLA/SL), meaning multi-dimensional printing* [25] Following the stereolithography process, additional methods were also added to the liquid raw material-based process arsenal *with digital light processing* (DLP) where a UV light projector (replacing the laser) while curing a whole layer at once, and the liquid container (vat) of the both processes was replaced by an inkjet printer setting in *polyjet printing*, also employing a UV light. This section covers the following processes with liquid raw materials:

Fig. 5.16   SonicLayer 4000—Hybrid additive/subtractive system [23]

- Stereolithography Process (SLA/SL)
- Digital Light Processing (DLP)
- PolyJet 3D Printing.

5.1.8   Stereolithography (SLA/SL)

The lead author was introduced to the 3D Systems' Viper SLA si machine in 2005 at RMU. He had become one of the operators of the machine and played that role for another 5 years. At the time, an annual service contract for the machine cost for universities about $10 K, replacing a crystal for manipulating the laser cost about $3 K, and the controller itself was around another $10 K. Price point for these machines were more than $200 K in their early existence—Fig. 5.17. Today, we can purchase an SLA machine with a small build volume for 2–3 hundred dollars.

The original SLA process (Viper) was equipped with a solid-state laser (UV laser) to harden UV curable liquid resin (photopolymer) [26]. The liquid resin was in monomer form but with the exposure to the heat of the laser, a chemical reaction took place facilitating the curing (polymerization) process solidifying the liquid material—Fig. 5.18. The laser solidified an area resembling a bullet pattern, and each layer had to adhere to its neighbors—Fig. 5.19a and b, by overcurving but not fully curing each layer. The laser started scanning each layer from its border, then produced a hatch pattern, filling in the gaps later as seen in Fig. 5.19c. The laser originated as Neodymium Yttrium Vanadate Nd:YVO4, turned infrared and green at the head of the machine, and UV at the vat (Class IIIb), requiring an elaborate system (Fig. 5.20). A secondary diode based red laser (Class

Fig. 5.17 3D Systems Viper si SLA machine

II) was also used in resin leveling in the machine. After the print is completed, it is raised and taken out of the vat, and cleaned in an alcohol bath (isopropanol) to remove its uncured resin coating and breakaway its softened supports. The cleaning operation is done often in a fume hood with UV filters covering its face, is the first step of post-processing. The cleaning can be also done in ultrasonic cleaners and other devices, if sterilization is needed—Fig. 5.21. The second step of post processing is post-curing with UV as shown in Fig. 5.22, earmolds for customized hearing aids are placed on the UV fluorescent lights in the device. This step cures the liquid pockets between bullet patterns shown in Fig. 5.19.

The lead author over the years utilized various SLA resins including Accura SI 40, (DSM) Somos resins like 11,120, and NanoTool, with added ceramic particles. He also used 3D Systems' Projet 1200 machine with wax material including composites development for printing resistors and capacitors employing carbon black and graphene—Fig. 5.23. Materials and low-cost machine development drove the SLA markets during their short history. The Projet 1200 (a micro SLA) cost much lower than its predecessors, but had a small build volume for fine printing. With the emergence of Formlabs, SLA process became truly affordable with a price tag less than $4 K, excluding its wash and post-cure stations—shown in Figs. 5.24 and 5.25. Finally, today, machine series like Any-cubic Photon or Elegoo can be purchased for a few hundred dollars, and the materials are available at much lower costs compared to the earlier days. Formlabs i.e. offers standard, mixing, dental, and engineering resins.

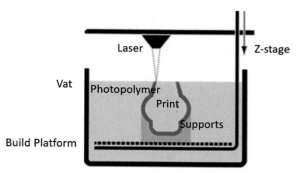

Fig. 5.18 SLA process [26]

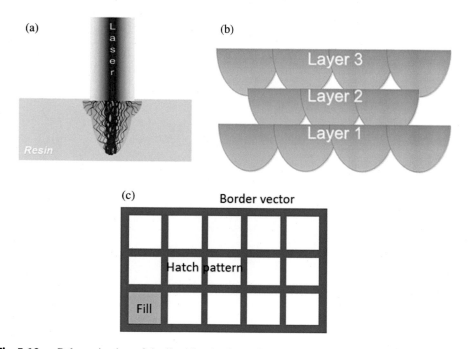

Fig. 5.19 **a** Polymerization of the liquid resin (heat affected zone) **b** Layer formation [26] **c** Layer scan pattern

New SLA machines (i.e. Formlabs, Anycubic Photon, Elegoo) are of inverted type, similar to the digital light processing machines. As the part is being printed, it is lifted, not lowered into the vat. This solves one of SLA's issues, vat's height limiting the height of the part being printed. As long as there is enough material in the vat, parts taller than the height of the vat can be printed with the new systems. Cross contamination issues are also partially eliminated since some of the materials come in cartridges and the scale

Fig. 5.20 SLA Viper laser

Fig. 5.21 Fume hood (left)
ultrasonic cleaner (right)

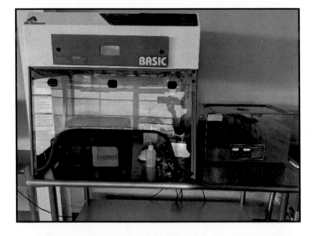

Fig. 5.22 SLA post-cure oven

(a) (b)

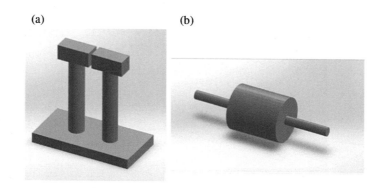

Fig. 5.23 Designs for SLA printing resin/graphene or carbon black **a** capacitors **b** resistors

(a) (b)

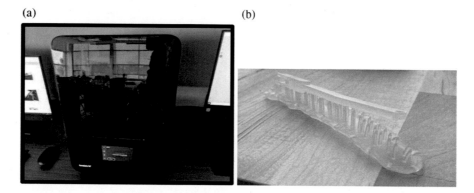

Fig. 5.24 **a** RMU Formlabs SLA printer **b** A parts made with a standard resin with its support being removed

of the machine also helps. Unlike the original SLA printed parts, new SLA parts are post-cured before the supports can be removed from the parts (Fig. 5.24b).

5.1.9 Digital Light Processing (DLP)

Besides employing similar resins used in SLA printing, digital light processing (DLP) method uses a digital projector to flash UV to the whole layer, curing all of its points simultaneously. The source reflects the light on a *digital micromirror devices (DMD)*, a dynamic masking device made of a microscopic-size mirror matrix on a semiconductor chip [27]. The part is also drawn out of the vat. This was the factor leading to the development of new inverted SLA machines. The DLP technology was developed by EnvisionTEC, a German company founded in 2002. At the time of its introduction to the

(a)   (b)

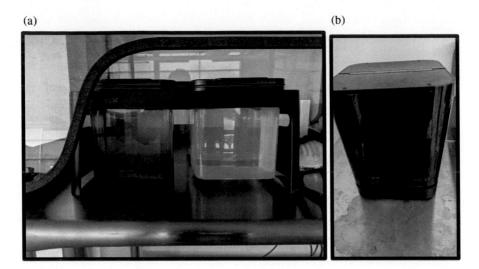

Fig. 5.25 Formlabs **a** wash stations **b** post-cure station

market, the DLP technology became a low-cost alternative to SLA machines since it was used for similar applications including customized earmold fabrication. Early 2021, DM acquired EnvisionTEC for $300 M [28].

The lead author was involved in development of a DLP machine in the time frame of 2014–2015. The work was initiated by using a timed Microsoft PowerPoint presentation along with a classroom projector (after removing its color wheel). Later on, an open-source software for DLP printers, Creation Workshop, became available [29]. After completing the printer, an optimization study was also conducted [30] (Figs. 5.26 and 5.27).

5.1.10 PolyJet Process

Polyjet process builds 3D forms by jetting layers of liquid photopolymer as low as 27 μm (approximately $0.001''$) simultaneously cured by a UV light [31]. The material is heated and jetted out in small droplets in a semi-viscous state to be cured. Gel-like support materials are removed by waterjet technology or automated methods dissolve the soluble ones. The technology was developed by Objet Geometries located in Rehovot, Israel in 1998. In 2009, the company released its first printer, but not long after it was merged with Stratasys (2012) [32].

PolyJet printers like Connex 350/500 can print with combinations of basic (digital) materials, resulting in a range of mechanical properties exhibiting rigid to flexible behavior including rubber-like, simulated, bio-compatible, and high temperature materials. It is

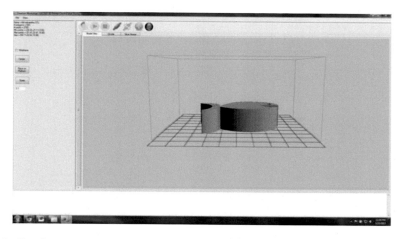

Fig. 5.26   Creation workshop [29]

Fig. 5.27   RMU DLP printer
[29]

(a)   (b)

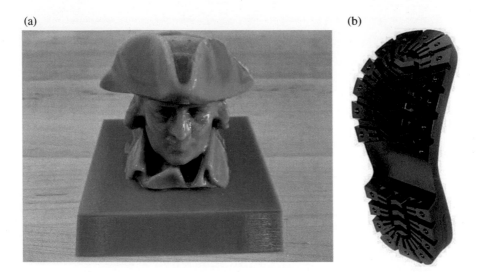

Fig. 5.28 **a** Objet30 Prime printed part with rigid opaque material **b** Objet Connex 500 printed shoe outsole

also possible to print with different materials on the same build tray in the same print job [33]. The prints can possibly have different (i.e. Shore A) hardness values at different areas of them (Fig. 5.28).

5.2   Powder Raw Material Based Processes

There is a wide range of 3D printing technologies based on powder-bed processing. Early ones included selective laser sintering (SLS) process and binder-jetting (was called 3D printing in early days of rapid prototyping). The need for powder processes lie in their ability of processing a multitude of materials (polymers, ceramics, and metals and their alloys) as well as making porous parts like bone implants, filter components, molds and vents. However, the main drawback of some of these systems include evacuation of unused powders after the print is complete. Powder based 3D printing technologies include:

- Binder-Jetting
- Multi Jet Fusion (MJF)
- Selective Laser Sintering (SLS)
- Direct Metal Laser Sintering (DMLS)
- Selective Laser Melting (SLM)
- Electron Beam Melting (EBM)
- Laser Engineered Net Shape (LENS®) Process.

5.2.1 Binder Jetting

Binder jetting is a process where an ink-jet printhead selectively deposits a liquid binder onto thin layers of powder particles [3]. The binder can work with foundry send, ceramics, metal or composite powders. The original binder-jetting process was invented by Ely Sachs in the late 1980s. Sachs, an MIT professor, also co-founded the company Desktop Metal [34]. Over the early years, two companies made their business based on the process, ExOne and Zcorp. While Prometal mainly focused on metal and sand mold printing, Zcorp used starches and gypsum powder. Both of these companies are owned by other companies today. 3D Systems acquired Zcorp in 2012, and recently Desktop Metal purchased ExOne. The lead author has had access to ExOne's Prometal RXD (prototype version of R1) and 3D Systems Projet 460+ machines—Fig. 5.29. He especially worked on stainless steel printing with the RXD machine, also employing ethylene glycol as its binder.

The binder jetting process has a multi-step workflow:

- The printer selectively drops binder on each layer of the powder being delivered from the powder reservoir. Drop-on-demand method is also used in this process. This step has four different sub steps:
  - Droplets are formed and released by the printhead.
  - Droplets impact the powder and break.
  - Droplets penetrate into the powder layers pores and form the internal core.
  - Internal cores experience a secondary growth.
- After the printing step is completed, the binder is removed by heating the print (debind). This gives the printed part some strength, then it can be removed from the powder bed—Fig. 5.30a.

Fig. 5.29 3D Systems Projet 460+ machine located in the RMU Learning Factory—a full color printer with gypsum as its particular media

(a)   (b)

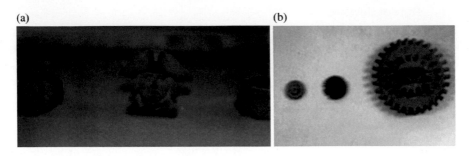

Fig. 5.30 a Small green parts printed with Prometal RXD **b** Small gears printed and sintered by Prometal equipment [35]

- The part is placed into the sintering furnace under vacuum first and inert gas (Argon Hydrogen mix for stainless steel) presence afterwards—Fig. 5.30b.
- After sintering, the part may be infiltrated with a dissimilar metal alloy (bronze for stainless steel) to bring it to full volume by filling its pores with capillary action. If the part is intended to be a porous one, this last step is not carried out.

Additional information on the binder jetting process, and its application to rapid tooling can be found in Chap. 7 of this book. The chapter covers ExOne's foundry sand mold printing process and its equipment in detail.

5.2.2   Multi Jet Fusion (MJF)

HP's *multi jet fusion (MJF)* process is equipped with an inkjet array to deliver selective areas with fusing and detailing agents across the powder bed, which are then fused by heating elements into a layer. The process was first introduced to the market in 2016 after a couple decades of research and development by HP Additive. After each layer, a new layer is placed on the previously printed layer and the process is repeated until the part is completed [36, 37].

The details of the process are given below [36]:

- The moveable build unit is placed into the printer.
- The material recoater moves across the build area to deposit a thin layer of the material.
- The printing/fusing carriage moves across the build area to preheat the material to a specific temperature to provide material consistency.
- An array of inkjet nozzles releases the agents onto the powder bed and heat lamp fuses the processed parts of the layer.
- After each layer is completed, the build unit retracts to create space for next layer of powder to be applied.

- The previous steps repeat until the part is completed.
- When the printing process is finished, the build unit contains both the printed part and unused powder. A separate processing station attached to the moveable unit is used to cool and unpack the printed part while recovering excess powder for later use.
- Bead blasting is employed to remove the loose powders covering the print. Air or water blasting can also be used for this step.
- Additional post-processing may also be required in the form of machining and manual sanding.

MJF process can handle two types of materials, rigid (Nylon (PA) 11 and (PA) 12, with or without the glass fill, PP) and flexible plastic ESTANE® 3D TPU M95A. Glass filled nylon material is reinforced with glass beads and results in higher stiffness and thermal stability than its standard version [36]. The process currently aims to address creating functional prototypes and small runs of end-use parts. It produces stronger parts (having tensile strength of maximum XY and Z 48 MPa/6,960 psi with the ASTM D638 method) at higher speeds compared to the selective laser sintering process [36].

5.2.3 Selective Laser Sintering (SLS)

Selective laser sintering (SLS) process, a support-less 3D printing method, was established by DTM Corporation located in Austin, TX in 1987. The company was acquired by 3D Systems in 2001. The SLS process is based on laser sintering powder materials in layers to make up a 3D shape. Earlier machines having CO_2 lasers (25–100 W), processed faster and had a good wavelength for their energy to be absorbed by the powders. They came with issues including hardware design difficulties, very high drive voltage requirements, and consequent overheating [38]. On the contrary, diode lasers are small, lightweight, and inexpensive, and have simple drive circuits. They can easily be cooled down with a heat sink since i.e. a 3W infrared diode laser may use 2.2 V and 3 Amps. However, diode lasers are only suitable for dark powders, and high-powered diode lasers comparable to their counterparts are much more expensive than those CO_2 lasers [38].

Original SLS systems had high initial investment costs ($250–500 K), also requiring peripherals and facility requirements as much as 200 sq. ft. These devices included an air handler and shifter, glass-bead finisher, furnace for infiltration of metal parts. Some systems weighed more than 6,000 lb. With the emergence of Formlabs SLS technology, a 30W Fuse 1 bundle with its cleaning station is currently about $28 K, and the machine itself is about 18 K (Fig. 5.31). The space requirements are also much smaller compared to the early times. However, these low-cost machines offer a limited number of materials at the moment (Nylon 12 (Fig. 5.32), Nylon 11, also with carbon fiber, and with glass fiber) even though new ones are under development. On the contrary, established industrial SLS machines can process a wide variety of materials including durable, flame retardant, impact resistant, rubber-like, wax for functional prototypes and end-use parts. These machines could also work with mixed metal powders (i.e. 420 stainless steel and bronze) coated with thermoplastic binders.

There are two separate workflows for the SLS process, one (i) for polymer powders, and other (ii) for thermoplastic-coated metal powders (Fig. 5.32):

(i)

- Laser sintering of the powders, after a new layer of powder is applied by a recoater.
- After completion of the prints, cooling and removal of the prints from the unused powder.

Fig. 5.31 Formlabs Fuse 1 SLS printer

Fig. 5.32 SLS process in-progress printing with Nylon 12 powder (dark areas were traced by the laser

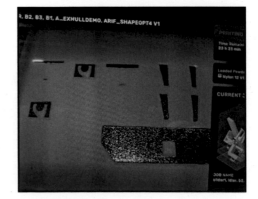

- Removal of the loose powder from the prints—Fig. 5.33, leading to finished part shown in Fig. 5.34.
- Conducting other post-processing if necessary.

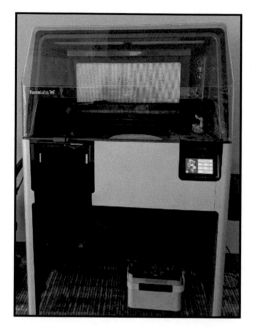

Fig. 5.33 Formlabs fuse 1 cleaning station

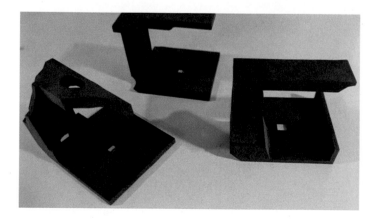

Fig. 5.34 Post-processed nylon 12 SLS prints employed in topology optimization work

(ii)

- Laser sintering of the powders, after a new layer of powder is applied by a recoater.
- After completion of the prints, cooling and removal of the prints from the unused powder.
- Debinding and sintering of the metal powders in a furnace.
- Infiltration of porous metal part by another molten metal.
- Conducting post-processing if necessary.

5.2.4   Direct Metal Selective Laser Sintering (DMLS)

The origins of the direct metal 3D printing dates back to 1994 when EOS introduced its EOSINT M250 machine. M250 was also the first *direct metal laser sintering (DMLS)* machine with dual 100 W lasers and a scan speed of 5 m/s. Over the years, EOS developed more powerful (ND: YAG) fiber lasers and today's DMLS machines (i.e. EOS M290 has a 400 W laser and 7.0 m/s scan speed) are able to melt the metal powders, fully achieving a density over 99%. Key companies like EOS, Concept Laser, and 3D Systems have been developing and offering metal 3D printing solutions fitting into the smart factory of the future. These solutions are modular, configurable, and offer a high-level automation to maximize efficiency while reducing dependency on manual labor [39].

The printing process workflow for an original EOS M series machine include:

- Recoater arm applying a layer of powder transferred from the powder reservoir.
- Exposing the layer to the laser source for the sintering process.
- Lowering the build platform.
- Moving the recoater arm to the powder reservoir.
- Raising the powder reservoir.
- Repeating the process steps above until the part is completed.
- After completion of depowdering (removal of the powder remaining on the build plate), the print needs to be removed from its supports using *wire electrical discharge machining (EDM)* or a band saw. Supports are also removed from the build plate.
- CNC machining can be done to improve the surface finish and bring the print to net shape as well as heat treating.

DMLS process can work with a variety of metals and metal alloys including: stainless steel, nickel steel, Sterling silver, gold, titanium, bronze, copper, brass, aluminum, cobalt chromium, and Inconel® [40]. Applications of this technology is also wide-spread across many different industries but the lead author has been following its applications locally in the dental engineering work including fabrication of cobalt chromium molybdenum (CoCrMo) copings and aligners.

5.2.5 Selective Laser Melting (SLM)

Selective laser melting (SLM) process employs a high-power laser (i.e. Yb-fiber optic laser up to 1 kW) to melt powder in a bed protected by an inert gas environment to fabricate 3d shapes. SLM workflow is similar to that of DMLS, with one difference—powder material is melted to fuse. Its roots date back to 1995, when the Fraunhofer Institute located in Aachen, Germany filed the first patent on the laser melting of metals [39]. It accounts for more than 80% of the metal 3D printer market today with a large number of manufacturer's offering machines in a wide range of sizes and features. It is also called with other names like *direct metal laser melting (DMLM), laser metal fusion (LMF), laser cusing, and laser powder bed fusion (LBF)*. These processes can handle ceramics, and dozens of metals and their alloys, but some of the most common ones can be listed as aluminum, cobalt chrome, copper, nickel-based alloys and Inconel®, stainless and tool steels, titanium alloys, and precious metals [41]. Its applications are most suited in the aerospace, automotive, construction, and jewelry industries.

5.2.6 Electron Beam Melting (EBM)

Electron beam melting (EBM) is a powder-bed fusion technology which can only work with conducting powders (i.e. titanium alloys and copper), limiting its material reach and applications (i.e. medical, aerospace, and automotive). It applies a high-power electron beam (4 kW) in a vacuum environment, in which the beam is steered by an electromagnet to selectively melt the powders. In 2000, Arcam AB located in Molndal, Sweden patented the EBM technology. However, it was not until 2002 its first machine, the S12, was released [39]. All3dp summarizes the process' benefits and drawbacks [42]:

- Benefits: ability to process high temperature, crack-prone, and reflective alloys, high part density, homogeneous microstructure and excellent mechanical properties, minimized need for heat treatment, 95–98% recyclable unused powder, faster than laser-based methods (scanning speeds of 1000 m/s and high build rates), fewer supports required compared to the laser-based methods.
- Drawbacks: limited print volume, limited material selection (including Titanium and its alloys, cobalt chrome, copper, nickel alloy, tool and stainless steels, tantalum and titanium tantalum alloy), expensive machine and materials, poor surface finish before post-processing, less detailed part designs compared to SLM, no-multi material option, cathodes needing to be replaced regularly, vacuum taking a long time to build up.

5.2.7   Laser Engineered Net Shape Process (LENS®)

The *laser engineered net shape (LENS®)* process was commercialized by Optomec in 1998. It is similar to WAAM technology (Fig. 5.13) which is also a direct energy deposition method [39]. Rather than using wire as feedstock, LENS® employs metal powders which are applied in its build chamber by the material deposition head. The built chamber's atmosphere includes purged argon keeping oxygen and moisture levels low, A high-power laser (400W to 3 kW) melts the powder into fully-dense structures. This method can be used in building back volume of worn mechanical (i.e. turbine, blisk airfoil and other aircraft engine) components using materials like titanium, stainless steel, and Inconel®. It can be 10 times faster and 5 times cheaper when compared to the other powder-bed based technologies. It can also produce excellent mechanical properties including fatigue properties. The method is compatible with automation and in hybrid manufacturing [43].

5.2.8   Recent Developments in Metal 3D Printing

Within the last decade, especially between the years of 2011–2018 before it was disrupted by the pandemic, metal AM sales exhibited exponential growth reaching almost $2.0B. The technology also has seen the following recent developments [39]:

- 2016: GE Purchases controlling shares of Arcam AB.
- 2017: Digital Alloys, a US-based start-up, patented its Joule 3D metal printing method and closes a $12.9 Series B funding round in 2018.
- 2018: HP introduces its Metal Jet 3D Printing systems for metals.
- 2019: Desktop Metal closes a $160 M Series E funding round.
- 2021: Desktop Metal acquires ExOne and EnvisionTEC.

Thus, as the new technologies become available and enter the market, they are also drawing attention, including financial support while the acquisitions and mergers are continued.

5.3   Other Processes

There have been many attempts in developing and commercializing new processes including the ones mentioned in the previous section. We can categorize some of these processes as *other* including *cold spray additive manufacturing*. It aims to create a continuous coating on a substrate by applying metal powders. The process can create a single-layer coating or accumulate many layers to build up a 3D form for a part [44]. Printing with pastes, plaster, clays, and concrete has a great potential too. Especially concrete 3D printing has become very popular due to its applications in the construction industry, but needs to overcome additional hurdles in scaling its ways. The following subsection focuses on printing with paste materials eliminating the need of debinding process step found in other processes.

5.3.1   RAPIDIA Process

RAPIDIA's 3D printing process is similar to Desktop Metal (DM)'s original Bound Deposition Modeling (BMD) method. The main differences between the two processes are (Fig. 5.35):

- While BMD uses rods made from powder metal such as stainless steel with wax and polymer additive, RAPIDIA employs a metal paste in a container. According to the company [45], its paste, mainly made with water, includes some amount of polymer, but it is solvent free.
- There is no need for debinding of RAPIDIA printed parts as it is needed for the DM's BMD method.
- The second step of the RAPIDIA process is the sintering which is similar to the third step of the BMD process.

In response to RAPIDIA's two-step process, Desktop Metal has recently announced that they have also eliminated their debinding step. More information on this can be found in Sect. 5.1.3.

Fig. 5.35 RAPIDIA printer and its sintering furnace [45]

Review Questions

1. What are the different classifications of 3D printers? Prepare a mind map for it.
2. What are the solid raw material based processes? Compare and contrast them by using the nature of the processes?
3. What are the liquid raw material based processes? Compare and contrast them by using the nature of the processes?
4. What are the powder raw material based processes? Compare and contrast them by using the nature of the processes?
5. Why are there two terms used for the same process, FDM and FFF? Explain briefly.
6. Study the composite material development in FDM/FFF and SLA processes including the lead author's work and summarize in two PowerPoint slides, one for each process.
7. What is the main drawback of the original SLA machines in terms of the parts they make?
8. Which process can mix materials and have different hardness values within the same print?
9. What makes the binder jetting process cumbersome?
10. Which equipment(s) can print in full color even though the prints may not be functional?
11. What is the main advantage of powder-based prints in terms of their applications? (Hint: Think about both the medical and industrial applications)
12. What is the method used in preventing corrosion during prints in multiple powder-based fusion methods?
13. What is the main limitation of the EBM process?

Research Questions

1. Research about the *Introduced Profiled Edge Lamination (PEL)* tooling method including its nature and advantages over other laminated tooling methods.
2. What are the other ways of preventing warping by using other Cura FFF process settings (Hint: Use the Ultimate Cura Tutorial as a reference)
3. Print at least 6 1-inch cubes and measure to see how accurate and repeatable they are. You can use the same FFF 3D printer and two different materials of your choice to see how they respond to the same process parameters.
4. Print at least 6 1-inch cubes and measure to see how accurate and repeatable they are. You can use any printer available to you and two different materials of your choice to see how they respond to the same process parameters.
5. Study the failure modes in the SLA/SL printing process making a list of them and their reasons.
6. What is the CLIP process? Prepare a two slide PowerPoint presentation about it.

7. Explain the methods of continuous and drop-on-demand (DOD) material jetting technology, comparing them to each other.
8. What is the new TPMS Infill employed in BMD? Elaborate briefly with a design example.
9. Prepare a mind map for metal 3D printing.
10. Prepare a mind map for polymer 3D printing.
11. Prepare a mind map for ceramics 3D printing
12. What is the Joule 3D printing process? Is it similar to other processes?
13. How does the Raise's new Hyper FFF printing process work?
14. How does the HP Metal Jet 3D printing process work?
15. Recently Markforged acquired a software company, Teton Simulation Software, for adding optimization to its software arsenal. What is the importance of Teton?
16. What are the top trends in 3D printing and AM today?
17. Conduct a literature review on the status of micro 3D printing. Summarize your findings in three PowerPoint slides.

Discussion Questions

1. What is the difference between utilizing sintering and melting in making metal parts? Which one do you think performs better in obtaining fully dense parts?
2. Discuss importance of preventing stress cracking in powder-bed technologies.
3. Discuss ceramic 3D printing and its applications.
4. Discuss the need for additional composite materials.
5. After reviewing a part to be 3D printed with FDM/FFF, please use the Purdue University scorecard (attached below [46]) to analyze its designs fit to 3D printing.
6. What makes the following print (spring) and other similar designs flexible even though it was printed with a rigid PLA material?

7. Compare and contrast metal AM with powder metallurgy methods in a similar fashion found in Chap. 1 where AM was compared and contrasted against CNC machining.
8. Discuss the potential for battery AM as the automotive companies go electric.
9. Automotive industries are already enjoying contributions from AM. Discuss if they should also build a whole vehicle's body using AM as Oak Ridge National Laboratories did—see the GROVE given below.

10. After studying the present state of the 3D printing and additive manufacturing realm, please discuss its future. (Hint: You can use Wohler's Reports as references)

Mark One	Complexity	Mark One	Functionality	Mark One	Material Removal	Mark One	Unsupported Features	Sum Across Rows		Totals
	note: Simple parts are inefficient for AM		note: AM parts are light and medium duty		note: Support structures ruin surface finish		note: Unsupported features will droop			
□ †	The Part is the same shape as common stock materials, or is completely 2D	□	Mating surfaces are bearing surfaces, or are expected to endure for 1000+ of cycles	□	The part is smaller than or the same size as the required support structure	□	There are long, unsupported features	0	X5	0
□ *	The part is mostly 2D and can be made in a mill or lathe without repositioning it in the clamp	□ *	Mating surfaces move significantly, experience large forces, or must endure 100-1000 cycles	□	There are small gaps that will require support structures	□	There are short, unsupported features	0	x4	0
□	The part can be made in a mill or lathe, but only after repositioning it in the clamp at least once	□	Mating surfaces move somewhat, experience moderate forces, or are expected to last 10-100 cycles	□	Internal cavities, channels, or holes do not have openings for removing materials	□	Overhang features have a slopped support	0	x3	0
□	The part curvature is complex (splines or arcs) for a machining operation such as a mill or lathe	□	Mating surfaces will move minimally, experience low forces, or are intended to endure 2-10 cycles	□	Material can be easily removed from internal cavities, channels, or holes	□	Overhanging features have a minimum of 45 deg support	0	x2	0
□	There are interior features or surface curvature is too complex to be machined	□	Surfaces are purely non-functional or experience virtually no cycles	□	There are no internal cavities, channels, or holes	□	Part is oriented so there are no overhanging features	0	x1	0

Mark One	Thin Features	Mark One	Stress Concentration	Mark One	Tolerances	Mark One	Geometric Exactness			+
	note: Thin features will almost always break		note: Interior corners must transition gradually		note: Mating parts should not be the same size		note: Large flat areas tend to warp			
□	Some Walls are less than 1/16" (1.5mm) thick	□	Interior corners have no chamfer, fillet, or rib	□	Hole or length dimensions are nominal	□	The part has large, flat surfaces or has a form that is important to be exact	0	x5	0
□	Walls are between 1/16" (1.5mm) and 1/8" (3mm) thick	□	Interior corners have chamfers, fillets, and/or rib	□	Hole or length tolerances are adjusted for shrinkage or fit	□	The part has medium-sized, flat surfaces, or forms that are or should be close to exact	0	x3	0
□	Walls are more than 1/8" (3mm) thick	□	Interior corners have generous chamfers, fillets, and/or ribs	□	Hole and length tolerances are considered or are not important	□	The part has small or no flat surfaces, or forms that need to be exact	0	x1	0

				Overall Total	0

	Starred Ratings			Total Score	
*	Consider a different manufacturing process			33-40	Needs redsign
†	Stongly consider a different manufacturing process			24-32	Consider redesign
				16-23	Moderate likelihood of success
				8-15	Higher likelihood of success

References

1. Non-planar 3D printing Simply Explained, located at: https://all3dp.com/2/non-planar-3d-printing-simply-explained/, accessed September 10, 2022.
2. Badger, P.D., Sirinterlikci, A., Yarmeak, G. (2016). Conversion of a 3-Axis Commercial Milling Machine into a 3D Printer, Journal of Engineering Technology, 33 (2), pp. 40-48.
3. The 7 categories of Additive Manufacturing, located at: https://www.lboro.ac.uk/about/the-seven-catogories-of-additive-manufacturing/, accessed September 10, 2022.

4. Nakagawa, T., Kunieda, M., and Liu, S. D. (1985). Laser Cut Sheet Laminated Forming Dies by Diffusion Bonding, Proceedings of the 25th International Machine Tool Design and Research Conference, U. of Birmingham, England, pp. 505–510.

5. Walczyk, D.F. and Hardt, D.H. (1998). Rapid tooling for sheet metal forming using profiled edge laminations-design principles and demonstration, Journal of Manufacturing Systems, 120 (4) , pp.746-754, https://doi.org/10.1115/1.2830215.

6. US5121329A, Apparatus and method for creating three-dimensional objects, 1992–06–09 Application granted and published.

7. Cura Settings Decoded—An Ultimate Cura Tutorial, located at: https://m.all3dp.com/basics, accessed September 10, 2022.

8. 3D Printing Supports—The Ultimate Guide, located at: https://m.all3dp.com/1/3d-printing-support-structures, accessed September 10, 2022.

9. RepRap—RepRap, located at: https://reprap.org/, accessed September 10, 2022.

10. Sirinterlikci, A., Badger, P., Hillwig, M., Tabassum, R., Schoonhoven, J., Kaczmarek, J. (2022). Development of Antimicrobial Polymer-Metal Composites via Additive Manufacturing Methods", 2022 Rapid +TCT Conference, Detroit, MI.

11. The Best Wood PLA Filaments of 2022 - All3DP, located at: https://all3dp.com/2/wood-filament-for-a-3d-printer-explained-compared/, accessed September 10, 2022.

12. Markforged Onyx One Datasheet, located at: https://s3.amazonaws.com/mf.product.doc.images/Datasheets/F-PR-4011.pdf, Markforged Onyx Pro Datasheet, accessed September 10, 2022.

13. EC2F, located at: https://www.raise3d.com/e2cf/, accessed September 10, 2022.

14. Zwier, M. P., Wits, W. W. (2016), Design for Additive Manufacturing: Automated Build Orientation Selection and Optimization, Procedia, 55, pp. 128-133, https://doi.org/10.1016/j.procir.2016.08.040.

15. Markforged Onyx Pro Datasheet, located at: https://s3.amazonaws.com/mf.product.doc.images/Datasheets/F-PR-4012.pdf, accessed September 10, 2022.

16. 3D Printing with Composites with Markforged, located at: https://markforged.com/resources/learn/design-for-additive-manufacturing-plastics-composites/3d-printing-composites-with-markforged, accessed September 10, 2022.

17. Deep Dive: Bound Metal Deposition (BMD), located at: https://www.desktopmetal.com/resources/deep-dive-bmd, accessed September 10, 2022.

18. BMD Design Guide, located at: https://www.desktopmetal.com/uploads/Desktop-Metal-BMD-Design-Guide.pdf, accessed September 10, 2022.

19. Desktop Metal launches Studio System eliminating solvent debind phase—TCT Magazine (February 2, 2021), located at: https://www.tctmagazine.com/additive-manufacturing-3d-printing-news/desktop-metal-studio-system-2-eliminating-debind-step/, accessed September 10, 2022.

20. Renganathan, S. (2018), A new approach to Wire-Arc based Additive Manufacturing Process, located at: https://www.linkedin.com/pulse/new-approach-wire-arc-based-additive-manufacturing-sriram-renganathan, accessed September 10, 2022.

21. Wikipedia LOM reference, located at: https://en.m.wikipedia.org/wiki/Laminated_object_manufacturing/, accessed September 10, 2022.

22. About Us - |Fabrisonic, located at: https://fabrisonic.com/about-us/, accessed September 10, 2022.

23. Technology, 3D Printing Metal Without Melting, located at: https://fabrisonic.com/technology/, accessed September 10, 2022.

24. History of Stereolithography, located at: https://protechasia.com/en/stereolithography/history-of-stereolithography, accessed September 10, 2022.

25. ASME Historic Mechanical Engineering Landmark Stereolithography, https://www.desktopme tal.com/resources/metal-finishing-for-3d-printed-parts/, accessed September 10, 2022.
26. 3D Systems, the Viper SLA System Operations Course, Hearing Aid Accounts.
27. SLA vs. DLP: Guide to Resin 3D Printers—Formlabs, located at: https://formlabs.com/blog/resin-3d-printer-comparison-sla-vs-dlp/, accessed September 10, 2022.
28. Desktop Metal buys fellow 3D printing company EnvisionTEC for $300M, located at: https://techcrunch.com/2021/01/15/desktop-metal-buys-fellow-3d-printing-company-envisiontec-for-300m/amp/, accessed September 10, 2022.
29. Sirinterlikci, A., Moran, K., Kremer, C., Barnes, B., Cosgrove, J., Colosimo, S. (2015) , A Capstone Project on Designing and Developing a DLP 3D Printer, ASEE Annual (American Society for Engineering Education) Conference and Exposition, Seattle, WA.
30. Erdem, E., Sirinterlikci, A Statistical Approach for Process Optimization of Digital Light Processing (DLP) 3D Printing Process, International Journal of Rapid Manufacturing, https://doi.org/10.1504/ijrapidm.2020.10019626.
31. PolyJet Parts On Demand—Stratasys, located at: https://www.stratasys.com/polyjet/, accessed September 10, 2022.
32. What is PolyJet technology and how does it work?—Trimech, located at: https://trimech.com/blog/what-is-polyjet-technology-and-how-does-it-work/, accessed September 10, 2022.
33. Connex 350/500, located at: https://support.stratasys.com/en/printers/polyjet-legacy/connex 350-500, accessed September 10, 2022.
34. Ely Sachs Is the Living Embodiment of 3-D Printing, located at https://www.forbes.com/sites/jimvinosk, accessed September 10, 2022.
35. Sirinterlikci, A. (2012), Fabricating Small Functional Parts with a 3D Metal Printer, SME Rapid/3D Imaging Conference. Atlanta, GA.
36. What is MJF (HP's Multi Jet Fusion) 3D Printing?, located at: https://www.hubs.com/knowle dgebase/what-is-multi-jet-fusion/, accessed September 10, 2022.
37. HP Multi Jet Fusion 3D Printing Technology—Powder 3D Printer |HP® Official Site, located at: https://www.hp.com/us-en/printers/3d-printers/products/multi-jet-technology.html/, accessed September 10, 2022.
38. DIY Selective Laser Sintering FAQ, located at: https://reprap.o/wiki/DIY_Selective_Laser_Sin tering_FAQ/, accessed September 10, 2022.
39. Metal 3D Printing: Where Are We Today?, located at: https://amfg.ai/2019/02/19/metal-3d-pri nting-where-are-we-today/amp/, accessed September 10, 2022.
40. An Introduction to Direct Metal Laser Sintering, located at: https://amfg.ai/2017/03/15/introd uction-direct-metal-laser-sintering/amp/, accessed September 10, 2022.
41. Selective Laser Melting (SLM 3D Printing)—The Ultimate Guide, located at: https://m.all3dp.com/1/selective-laser-melting-guide/, accessed September 10, 2022.
42. Electron Beam Melting (EBM 3D Printing)—The Complete Guide, located at: https://m.all 3dp.com/2/electron-beam-melting-ebm-3d-printing-simply-explained/, accessed September 10, 2022.
43. LENS®, LENS Directed Energy Deposition (DED) 3D Printed Metal Technology, located at: https://optomec.com/3d-printed-metals/lens-technology/, accessed September 10, 2022.
44. Cold Spray Additive Manufacturing, located at: https://www.ctc.com/, accessed September 10, 2022.

45. Technology—Rapidia—Fast & Simple Metal 3D Printing, located at: https://www.rapidia.com/, accessed September 10, 2022.
46. Booth, J.W., Alperovich, J., Reid, T.N., Ramani, K. (2016), The Design for Additive Manufacturing Worksheet, Proceedings of the ASME International Design Engineering Technical Conferences and Computers and Information in Engineering Conference, IDETC/CIE 2016, Charlotte, NC.

3D Printing Post Processing

6

Quality including surface quality of 3D printed or AM parts are mainly determined by the material selection, process chosen and its parameters as well as the design of the parts, as 3D printing and AM technology is challenging the conventional methods to replace them. Post-processing of 3D printed or AM parts is also very critical to the outcome of the overall process, including resulting surface quality, total cycle time and cost. Even though this chapter is designated to post-processing, Chap. 5 also includes a sizable amount of information on the subject. Additional information is spread throughout the book.

This chapter covers removal of support structures and their cleaning, and consequent finishing processes while focusing on the FDM/FFF and SLS processes.

6.1 Removal of Supports and Cleaning

Support structures determine if a 3D printing or AM process can successfully achieve its aim. They are also vital to the quality of the surface of the part. Thus, support design needs to be given special attention. (Fig. 6.1).

Support removal (Fig. 6.2) is the most basic form of post-processing of 3D printed or AM parts. Usually, support removal doesn't require much effort, unless there are supports in tight corners or other hard-to-reach places. Depending on what they're made of, supports can be insoluble or soluble (which are able to be dissolved in water or another liquid).

Insoluble supports are made from the same material as the main part. FDM/FFF 3D printers with a single extruder can only use this type of support, as the part and its supports will be printed from the same spool of filament. Removing insoluble supports is usually done by snapping them with your fingers (breakaway type supports including new accordion ones, Fig. 6.2, left) or cutting them with a pair of pliers.

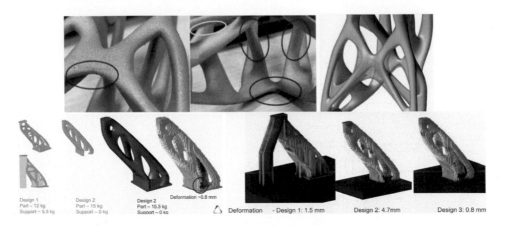

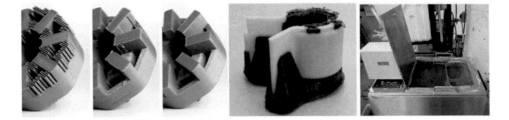

Fig. 6.1 Poor surface quality due to overhangs below the threshold can be eliminated by reducing overhangs using altair inspire 3D print simulation for metal powder bed fusion 3D printing [1]

Fig. 6.2 Original print with support attached: poor support removal and good support removal (left picture- left to right) The brown support material surrounding this AM part must be removed before the part can be used (right picture- left). A solvent bath is one way in which support material can be removed ((right picture- right) (Image courtesy of Hubs [4])

If a dual-extruder 3D printer is used, users will be able to apply soluble supports (Fig. 6.2, middle and right). Whereas insoluble supports can be very tricky to remove in places that are hard to reach, soluble supports can simply be dissolved by soaking the part in either water or another liquid, leaving little to no marking or residue. However, this process may take time even with the application of heat, ultrasonic waves, and other chemicals as catalyzers.

6.2   Finishing Processes

Once the part is removed from the 3D printer and is free of support material, it is ready for finishing. Post-processing encompasses a variety of finishing processes that occur after the supports are removed. These processes include:

- Abrasive finishing, such as using sandpaper to smooth rough areas or polishing the part's surface.
- Coating application, which involves applying surface treatments, such as a sealant to improve the part's mechanical properties.
- Post-curing, which uses ultraviolet (UV) or visible radiation to completely harden a part, and is often necessary for parts built with photopolymerization.
- Heat treating, which applies to AM parts made from metal, may be required as a final step to strengthen the part or relieve internal stresses that develop during the process.
- After sawing of metal AM supports, parts may require machining for finishing including surface finishing.

Different AM processes and materials will require different post-processing steps. Some parts may require little or no post-processing, while others may require extensive work. A part that must have very tight tolerances. for example, might require more finishing than a part with less specific dimensions. A part that must have smooth surfaces also requires polishing.

6.2.1   Post Process for Material Extrusion (FDM/FFF)

Some of the most widely used AM systems use a method known as material extrusion as mentioned in Chap. 5, which is sometimes called fused deposition modeling (FDM) or Fused Filament Fabrication (FFF). Material extrusion systems, as shown in the figure below 9 (Fig. 6.2, left), use a nozzle to dispense material onto a support. Most material extrusion processes use a polymer or thermoplastic material in the form of a solid filament. During operation, the filament is softened, either by heat or chemical reactions and then pushed through the extrusion head. The material solidifies and joins with additional layers to form the part. FDM 3D printing is best suited for cost effective prototypes produced with short lead time. Layer lines are generally present on FDM prints making post processing an important step if a smooth surface is required. Some post processing methods can also add strength to prints helping to mitigate the *anisotropic behavior* of FDM parts (*where material will exhibit better strength on the XY plane compared to that of the Z-axis*).

If a breakaway support material was used, the support material is easily removed manually by breaking it away from the final part. If a soluble support material was used, it is

Fig. 6.3 If a breakaway Support material was used, the support material is easily removed manually by breaking it away from the final part. (left) If a Soluble Support material was used, it is removed in a cleaner tank, using a vibrated water-based detergent solution. (right)

removed in a cleaner tank, using a vibrated hot water-based detergent solution (sodium hydroxide for ABS-like support materials containing biodegradable materials (Fig. 6.3, right). With the use of the hands-free soluble support, models with moving parts can be printed, assembled, and be fully functional once the support is washed away. Soluble support is also best for printing small intricate features, where the support can easily be dissolved away without stressing the part.

Another important advantage is the option to create products with moving parts, or small features, using Soluble support. This hands-free support removal solution saves time and costs associated with other systems that require manual labor to physically remove the support material (Fig. 6.4).

Standard dissolvable support materials are removed from a print by placing the print in a bath of the appropriate solvent until the support material dissolves. The support is typically printed in:

- HIPS (usually associated with ABS) (High Impact Polystyrene)
- PVA (usually associated with PLA) (Polyvinyl Alcohol)
- HydroFill [2].

Fig. 6.4 An FDM printed assembly exhibited after support removal

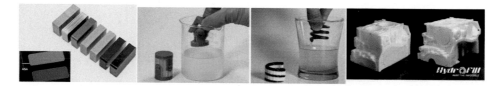

Fig. 6.5 Post-processed FDM prints (left picture- from left to right): Cold welding, gap filling, unprocessed, sanded, polished, painted and epoxy coated. Standard dissolvable support materials are removed from a print by placing the print in a bath of the appropriate solvent until the support material dissolves. (right)

Glass storage containers, like a mason jar, make excellent vessels for dissolving with Limonene. For dissolving in water, any non-porous container will work. For HIPS/ABS prints, a bath in a 1:1 ratio of limonene and isopropyl alcohol (IPA) works very well for rapid support removal [3].

Many other support materials, such as PVA (used with PLA) and HydroFill (PLA and ABS), simply dissolve in plain water [3].

FDM prints parts using production-grade thermoplastics that can be sanded, drilled, glued and painted just like any plastic part. FDM parts can be finished and painted to meet the cosmetic requirements of nearly any application. Virtually any tough, durable, and strong part can be finished to have a smooth, paint-ready surface. Smoothing surfaces can be done by sanding, melting the thermoplastics, or polishing as well as etching (i.e. employing acetone for ABS parts) (Fig. 6.5).

After supports are removed or dissolved, sanding can be done to smooth the part and remove any obvious blemishes, such as blobs or support marks (Fig. 6.6). Especially with FDM parts, where layers are easily visible, it's important to sand the parts in a circular motion. If parts are sanded parallel or perpendicular to the layers, the visual appearance of the parts could be ruined. The starting grit of sandpaper depends on the layer height and print quality; for layer heights of 200 microns and lower, or prints without blemishes, sanding can be started with 150 grit. Sanding should proceed up to 2000 grit (fine), following common sanding graduations (one approach is to go from 220 grit to 400 grit, to 600 grit, to 1000 grit and finally 2000 grit) (Fig. 6.7). It is recommended to wet sand the print from start to finish, to prevent friction and heat build-up from damaging the part and keep the sandpaper clean. The print should be cleaned with a toothbrush and soapy water. Sanding again should always be done in small circular motions evenly across the surface of the part. If the part discolors, or if there are many small scratches from sanding, a heat gun can be used to gently warm the print and soften the surface enough to "relax" some of the defects. In addition to sanding, i.e. a chemical like acetone can be used in spray form on ABS to smoothen part's surfaces.

3D printed parts can be bonded together (Fig. 6.5) to grow beyond the build envelope (i.e. ABS parts with acetone). For parts too large to fit on a single build, for faster job

(a)   (b)

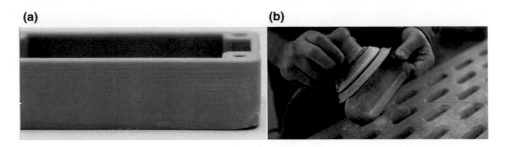

Fig. 6.6 **a** A sanded gray ABS print **b** Sanding-FDM Polishing

(a)   (b)

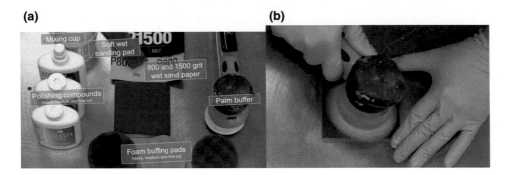

Fig. 6.7 **a** Sanding-FDM Polishing a gray PLA **b** FDM print spray painted blue

builds with less support material, or for parts with finer features, sectioning and bonding FDM parts is a great solution.

Once the print is properly sanded (only need to go up to 600 grit for painting), the print can be primed. Priming should be done in two coats, using an aerosol primer. The primed surface should be buffed and polished. Once priming is complete (Fig. 6.7b), painting can be done with artist's acrylic paints and brushes, but the use of an airbrush or aerosol will provide a smoother surface finish. Acetone is again well-known for its abilities to smooth ABS (Fig. 6.5). It produces professional results with attention to detail and some practice. It also allows for complete flexibility of the visual appearance of the final product, independent of the material/color the object was originally printed in. Acetone is highly flammable and can explode. Therefore, proper precautions should be followed while carrying out either of these processes. Even just the fumes can be harmful if inhaled, causing irritation and other negative effects. Post processing on 3D printed parts should be done in a well-ventilated area with proper PPE (personal protective equipment) such as gloves, safety goggles, and a mask.

Polishing is a post-processing technique that is used to achieve the smoothest possible surface. Polishing 3D prints can be done with plastic polishers and tools available at

(a)   **(b)**

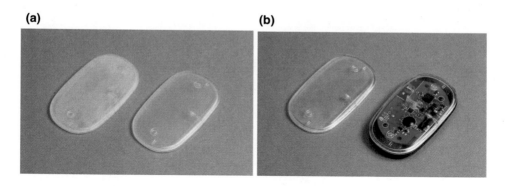

Fig. 6.8 Wet sanded (**a**-left) and mineral oil finish (**a**-right). Spray paint (clear UV protective acrylic) (**b**-left) and polished to clear transparent finish (**b**-right) (Image courtesy of Hubs [4])

almost every hardware store (Fig. 6.7a) A microfiber cloth and a plastic polisher can be used. A dremel tool can make this process easier.

Before a part can be polished, it needs to be sanded properly, finishing with the finest sandpaper. After sanding, rinse your parts and make sure there are no particles left. If you're using a cloth, apply the polisher to the sanded part and move the cloth in a circular motion until you're satisfied with the result. A Dremel buffer will do much of the work, but the user needs to make sure that it's moving evenly along the surface. It produces smooth and mirror-like surfaces and it is relatively inexpensive. However dimensional accuracy can be compromised, thus moderate skill-level is required (Fig. 6.8).

Electroplating is a process that adds a metal coating to other metals or parts with a conductive surface (please also see Chap. 7 for information on rapid tooling methods). It's a great post-processing technique that can significantly improve both strength and visual appearance (Fig. 6.9). This metal coating can be both decorative and functional. It gives the appearance of production metal or provides a hard, wear-resistant surface with reflective properties. Electroplated parts also have improved mechanical properties. If a user has taken advantage of the different build styles to print a light, inexpensive model very quickly, they can use one of the various sealing methods validated for FDM parts. Sealing makes it possible to benefit from a cost-saving printing method and have a liquid-holding or an airtight sealed product (Fig. 6.9). The RMU DLP machine had a vat with Teflon® coating to hold the resin of the process, as presented in Chap. 5.

Epoxy coating 3D prints improves strength, but it can also seal porous parts of the print and act as a great protective layer overall. The epoxy coat consists of two different chemicals: the epoxy resin itself and a hardener. Once the print is sanded (sanding first will produce better results), the print can be fully cleaned with a tack cloth. The appropriate ratio of resin to hardener as specified on the instructions for the resin should be mixed, ensuring everything is measured precisely. Epoxy resins are exothermic when mixed, so glass containers and containers composed of materials with low melting points

Fig. 6.9 A smoothed black ABS hemi-sphere print (Image courtesy of Hubs [4])

should be avoided. Very thin layer of epoxy will not impact the tolerances of the print all that greatly (unless the print is sanded first). It will provide an outer protective "shell" around the print (Fig. 6.9).

Metal-plating can be done using electroplating at home, or a professional shop. Proper metal-plating requires a strong knowledge of materials, and what can be done at home or a laboratory is limited in comparison to what a professional shop can achieve. For superior finishes, and a wider range of plating options including chroming, again utilizing a professional shop is the best option. A plated metal shell increases the strength of the plastic part, which greatly broadens potential applications and uses of the print. The outer metal coating is very thin, so tolerances can be tightly held if the plating is done properly. It can produce a beautiful surface finish, which if done properly, will not look like a 3D printed polymer object (Fig. 6.10).

One of the main strong points of the FDM process is its use of real thermoplastics. Thermoplastics are the same materials found in many of the common plastic items surrounding us, including television remotes, calculators, LEGO™, and many other everyday consumer goods. It enables printing of models that are accurate, durable, and stable.

Fig. 6.10 A black ABS print showing half coated with epoxy and half unprocessed (left) electroplated part (right) (Image courtesy of Hubs [4])

Working with materials with known properties and specified operating limits, makes them predictable, providing the option to select the best solution for the product.

6.2.2 Post Processing of SLS Parts

Parts produced with SLS 3D printing have a high level of accuracy, good strength and are often used as functional parts. In SLS, a laser selectively sinters polymer powder particles, fusing them together and building a part layer-by-layer. SLS produces functional plastic parts with isotropic mechanical properties that can be used for detailed prototyping or end-use low-volume production. Because of the nature of the powder based fusion process, SLS printed parts have a powdery, grainy finish. Post-processing of SLS parts is common practice with a range of techniques and finishes available (Fig. 6.11). Coatings are also regularly added to SLS parts to improve the performance. Furthermore, a functional coating can sometimes help to compensate for the lack of feasible material grades for SLS.

Parts are removed from the build chamber and all powder is removed from the part with compressed air in an SLS cleaning station (Chap. 5). The surface is then also cleaned via plastic bead blasting to remove any non-sintered powder sticking to the surface. This finish is inherently rough, like a medium grit sandpaper (satin-like matte finish that is slightly grainy). This is also the best surface preparation for painting or lacquering. All SLS parts come with this standard finish (unless otherwise specified). Good accuracy can be obtained since overall geometry is not altered.

For a smoother surface texture, Nylon SLS parts can be polished in media tumblers or vibro machines (Fig. 6.12). A tumbler that contains small ceramic chips that vibrate against the object gradually erodes the outer surface down to a polished finish. As a result, this process does have a small effect on part dimensions and results in rounding sharp edges. Thumbling is not recommended for parts with fine details and intricate features.

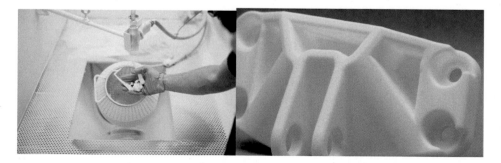

Fig. 6.11 Powder removal of SLS parts (left). The standard finish on an SLS 3D printed part (right) (Image courtesy of Hubs [4])

Fig. 6.12 The standard finish on an SLS 3D printed part (left). SLS parts being tumble/vibro polished. (center). A glossy spray paint finish on an SLS part (right). (image courtesy of hubs [4])

However, it provides an excellent smooth surface and removes sharp edges for parts with simpler designs.

The fastest most cost-effective method to color SLS prints is via a dye process. The porosity of SLS parts makes them ideal for dyeing (Fig. 6.12). The part is immersed in a hot color bath with a large range of colors available. Using a color bath ensures full coverage of all internal and external surfaces. Typically the dye only penetrates the part down to a depth of around 0.5 mm meaning continued wear to the surface will expose the original powder color. SLS parts can be spray painted. SLS parts can also be coated with a lacquer (varnish or clear coat) (Fig. 6.12). Via lacquering it is possible to obtain various finishes, such as high gloss or a metallic sheen. Lacquer coatings can improve wear resistance, surface hardness, water tightness and limit marks and smudges on the surface of the part. Due to the porous nature of SLS it is recommended that 4−5 very thin coats are applied to achieve a final finish rather than 1 thick coat. This results in a faster drying time and reduces the likelihood of the paint or lacquer running. Lacquer coating can improve mechanical properties and results in a glossy smooth colored or clear surface. It also improves UV protection.

Review Questions

1. What constitutes post-processing in 3D printing and AM?
2. What is the impact of not-properly done post processing 3D printed/AM parts?
3. Describe support removal in multiple processes. (Hint: Compare and contrast one liquid-, one solid-, and one powder-based process).
4. What is a "breakaway" support?
5. What are "water-soluble" and "biodegradable" supports?
6. List and define major finishing methods available for 3D printing/AM.
7. How are some of the finishing methods covered in this chapter employed for fabrication rapid tooling?
8. How are some of the finishing methods covered in this chapter employed for fabrication of functional parts or prototypes?
9. Explain the sanding process in an example.
10. Explain the painting process in an example.

11. What is cold welding? Explain with an example.
12. How can you convert an FDM/FFF print to a water-proof container? Briefly elaborate.

Research Questions

1. Prepare a mind map covering all post-processing methods including the ones for support removal, and finishing.
2. Research how polishing is handled in metal AM.
3. Summarize how polishing is carried out for ceramic 3D printing/AM in a two slide PowerPoint presentation.
4. Determine how old and new SLA post-processing methods differ.

Discussion Questions

1. Discuss whether FFF and FDM post-processing work-flow is any different or not.
2. Discuss the differences of SLS and MJF post-processing workflows.
3. 3D Systems' FinePoint™ supports had a complex but effective structure as they made fine point contacts with the print. Based on the following Fig. [5], please discuss their structure, including their components and relevant functionalities.
4. Discuss the secret costs of post-processing after reviewing Chap. 10.

References

1. Simulation driven design for additive manufacturing, located at: https://www.altair.com/resource/simulation-driven-design-for-additive-manufacturing?lang=en. Accessed September 1, 2022.
2. Post-processing-SLE-and-MJF-printed-parts, located at: https://www.hubs.com/knowledge-base/post-processing-sls-and-mjf-printed-parts/. Accessed September 1, 2022.
3. Water soluble support – 3D print, located at: https://airwolf3d.com/shop/water-soluble-support-3d-print/. accessed September 1, 2022.
4. FDM 3d-printing post-processing an overview for-beginners, located at: https://all3dp.com/2/fdm-3d-printing-post-processing-an-overview-for-beginners/, accessed September 1, 2022.
5. 3D Systems, the Viper SLA System Operations Course, Hearing Aid Accounts.

Applications of 3D Printing

7

This chapter presents uses of 3D printing, industrial and non-industrial. As mentioned earlier, industrial uses of 3D printing can be categorized as rapid prototyping, rapid tooling, and additive (rapid) manufacturing, applicable for many industrial applications. Non-industrial uses of 3D printing include a wide variety of areas such as food, medicine, forensics, anthropology, biology, and arts, just to name a few, and also include many one-of custom works, or items in small quantities.

7.1 Rapid Prototyping

As mentioned in Chap. 1, similar to any engineering design and development process, the product design and development effort *starts with planning and concept development, followed by system-level and detail design, and completed with testing and refinement, leading to production ramp-up (warm-up)* [1]. Virtual prototyping using CAD/FEA (Fig. 7.1) is employed early in the process in the detail design stage to detect design flaws as well as help estimate materials requirements for production.

With the testing and refinement stage, product developers fabricate and evaluate multiple physical prototypes of their design as they progress through the alpha and beta prototyping steps. Physical prototypes can be made via four ways, assuming that the materials used in them are already in pre-form:

- Subtractive means—machining/CNC machining
- Additive means—3D printing
- Formative means—casting/molding/forging/other shape forming processes
- Hybrid means—a combination of at least two methods given above.

A. Sirinterlikci and Y. Ertekin, *A Comprehensive Approach to Digital Manufacturing,* Synthesis Lectures on Mechanical Engineering, https://doi.org/10.1007/978-3-031-25354-6_7

Fig. 7.1 Results of
mechanical loading of
lower-limb prosthetic sockets
via CAE/FEA [2]

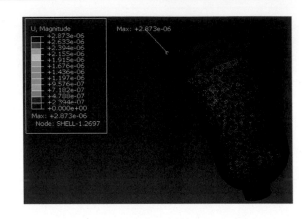

Role of 3D printing in rapid making of prototypes has increased greatly since it offers multiple advantages over its main alternative, the CNC machining. These advantages include having: much shorter lead times (i.e. hours versus days, days versus weeks), the ability to make almost any geometry possible, significantly less waste as well as no requirement of work-holding, other than the need of support structures which are also 3D printed along with the parts. With further developments in processes, process controls, and consequent quality as well as availability of a wide range of materials, 3D printing has become a major force for rapid prototyping. It can offer thorough *checking of form, fit, and function (3F"s)* in quick turnaround and affordable settings. The 3F"s of prototyping can be further explained with the following examples:

- Form check: 3D printed forms of parts or products give developers (Fig. 7.2) and their customers advantage in design presentations and reviews, as long as they are accurately printed.
- Fit check: Assembly of 3D printed parts presents an opportunity for checking parts" fit to the others in the assembly (Fig. 7.3).
- Function check: Even during the early times of 3D printing history, engineers found ways to print functional prototypes that can be used in testing the functional effectiveness of their design, such as in the case of 3D printed exhaust manifolds for internal combustion engines. Nowadays, with the availability of a wide variety of processes, process variables, and materials, the struggles of the early days look only a thing of a very distant past. Following figure (Fig. 7.4) illustrates checking of strength for 3D printed lower-limb prosthetic sockets.

In conclusion, rapid prototyping enables product manufacturers to become competitive, by facilitating rapid response times to the market with better quality and low-cost products since they can catch their mistakes early and improve not only the product designs, but also the roles of associated processes and equipment.

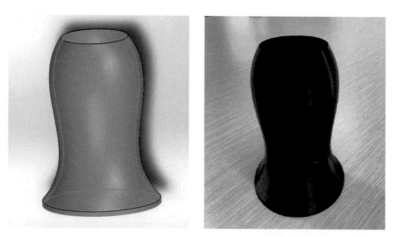

Fig. 7.2 A light shade (diffuser) design and its prototype for review of form

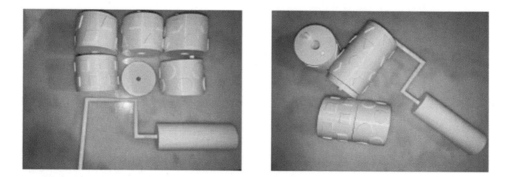

Fig. 7.3 Checking fit of 3D printed parts assembled in a modular toy design prototype for pressing patterns on clay or playdough

7.2 Rapid Tooling

Rapid tooling refers to rapid manufacturing of molds, dies, cutting tools (cutters), jigs/fixtures and other tooling components in short lead times. Rapid tooling lead times are much shorter than the conventional methods of tool making. It can be aimed for making either prototype or actual production tooling. Molds/cores made by rapid tooling can be used in injection molding or similar processes including sand casting. Tooling plates made by this type of effort are employed for the lost-foam casting process, and a similar approach can be taken for printing of porous vents for die-casting. In addition, 3D-printed cutting tools, specifically cutter inserts, are utilized for machining and especially in CNC machining along with 3D printed jigs/fixtures.

Fig. 7.4 Mechanical testing set-ups for static strength/compression test of 3D printed lower-limb prosthetic sockets [3]

There are two rapid tooling methods—*direct and indirect*. Direct rapid tooling is when the tooling component is designed in the CAD environment and printed for use as opposed to indirect rapid tooling where the tool engineers design patterns for mold making, then print those patterns, and finally manually make the molds around them (Fig. 7.5). Details of these methods and their examples are given in the following sections. If a 3D printed pattern is used in making all of the molds associated with the required work, the pattern is referred to as a *master pattern*.

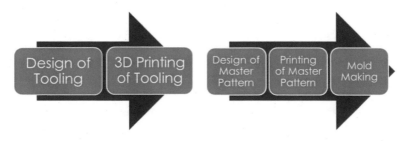

Fig. 7.5 Direct and indirect rapid tooling process flows

7.2.1 Direct and Indirect Rapid Tooling

In the early days of 3D printing, the processes were not as accurate and reliable as they are today. Thus, the improvements in the processes has paved the way for better rapid tooling over the years, along with help from materials development. Figure 7.6 presents a hierarchical chart of 3D printing-based rapid tool making, with its major segments direct, and indirect with soft and hard tooling. This chart is by no means fully comprehensive but offers a good summary of what major rapid prototyping methods exist or existed.

Fig. 7.6 A hierarchical chart for different direct rapid tooling methods

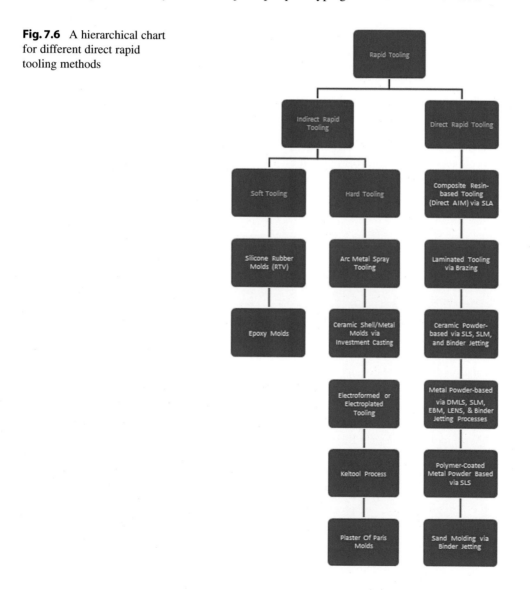

Table 7.1 Pros and cons of direct rapid tooling [4]

Pros	Cons
Faster production and shortened lead times	Often not as robust as prototypes made via indirect rapid tooling
Includes fewer process steps	Mold/tool making without patterns may introduce errors or discrepancies
May require fewer resources	Mold/tool making process needs to be repeated with tool failure
Can produce multiple prototypes from a single mold or tool	May not be appropriate for complex designs or materials
Extremely flexible and productive facilitating making of multiple versions of multiple molds/tooling	Could result in higher development costs due to selection of certain 3D printing technologies and multiple iterations

Direct rapid tooling can be accomplished by utilizing 3D printing technologies that can print stronger, harder materials with high temperature endurance when compared to those of indirect rapid tooling. However, they usually lack the performance of similar tooling made with conventional ways. Thus, direct rapid tooling is mainly suitable for short-run production, i.e. a few thousand parts. The Pacific Research Laboratories Engineering Department offers an excellent guide (Table 7.1) on direct rapid tooling and its role in making prototypes or parts in general [4].

Direct rapid tooling applications include but not limited to:

- Composite resin/ceramic or metal-based tool making for injection molding process via SLA process [5]
- Sheet metal tooling laminated with brazing or diffusion bonding [6]
- Ceramic powder-based tooling printed with a variety of methods including SLS, SLM, and Binder Jetting for a variety of manufacturing processes. This type of tooling may also include porous components like vents for die-casting.
- Metal powder-based tooling printed with a variety of methods including DMLS, SLM, EBM, LENS, and Binder-Jetting for a variety of manufacturing processes
- Polymer-coated metal powder printed with SLS for injection molding process
- Sand mold making via Binder-Jetting.

Indirect rapid tooling efforts can produce both soft and hard tooling, and often are associated with prototyping, and not with actual production. It is a good option for testing different materials and processes employing the same master pattern [4]. It can also be used in making a very small number of actual end-use parts. The Pacific Research Laboratories Engineering Department offers an excellent guide (Table 7.2) on using direct rapid tooling for making prototypes [4].

Table 7.2 Pros and cons of indirect rapid tooling [4]

Pros	Cons
The master pattern is durable and rarely gets damaged during tool making	Has longer lead times compared to direct rapid tooling
Will likely to invest in one master pattern unless the design changes over time	Involves intermediate process steps that could increase the costs
Can make either hard (for complex designs) or soft tools (for simple and cost-effective designs)	Not always a good option if the design changes significantly during the prototyping (testing and refinement) stage
Less variation amongst multiple copies of certain tools and molds, as they are all based on the same master pattern	May require higher quality materials to make a robust master pattern
Ideal for experimenting with different materials and printing technologies	Not always necessary for simple designs that don"t require a high dimension precision or accuracy

Indirect rapid tooling applications include but not limited to:

- Room Temperature Vulcanization (RTV) silicone-rubber mold making for polymer and low melting temperature range metal alloy casting processes [7]
- Epoxy resin mold making for a variety of accessories including jewelry, hair clips, or candles
- Arc metal spray tooling for injection molding process
- Ceramic shell tool making for investment casting and possibly consequent injection molding insert fabrication
- Electroformed tool making for slush molding an instrument panel skin [8]
- Electroplated tool making for injection molding process
- Keltool™™ process for making injection molding and die casting cores and inserts
- Plaster of Paris mold making for slip casting
- Lost PLA or wax for making metal castings from PLA/vax pattern coated with plaster.

Following sections include multiple case studies for direct and indirect rapid tool making. Additional information is covered by end of chapter questions and activities.

7.2.1.1 Direct Rapid Tooling of Injection Molding Inserts

This subsection is dedicated to the direct rapid tooling of injection molding inserts. As illustrated in Fig. 7.7, this type of molding process involves injection of liquefied polymers (originally in bead or pellet form) into temperature-controlled molds under high pressure. Once parts are solidified and cooled, the molds are opened and the parts are ejected. Figure 7.7 includes a simple two-piece mold with ejector (left) and cover (right) halves. A simpler mold insert and its associated part was designed, and the insert was 3D printed

Fig. 7.7 Injection molding
process [10]

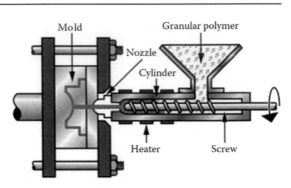

via the SLA (Viper) process by the lead author [5]. This insert (Fig. 7.8a) was made
from (DSM) Somos NanoTool material, a liquid epoxy resin filled with noncrystalline
ceramic nanoparticles. This material allowed printing of strong, stiff, high-temperature
resistant composite parts on an SLA machine for wind tunnel applications. It also permit-
ted faster processing, and exhibited superior sidewall quality (reducing finishing times)
and excellent detail resolution when compared to other SLA materials at the time [9].
The mold insert was attached to the ejector side of the molding, but lacked the holes for
ejector pins. Thus, the parts were run in the machine in a semi-automatic cycle setting to
assure the manual release of the parts. More than 100 parts in polycarbonate (PC) were
molded (Fig. 7.8b). Since the NanoTool material is thermally sensitive, it will continue
to be impacted, even in solid form, by the heat input of the molding process during each
cycle (shot) leading to aging of the mold material. Thus, the effort was intended as an
experimentation for a short-run direct rapid tooling. The cost of the tool insert was much
more economical compared to similar metal inserts (costing at least a few thousand dollars) even though the NanoTool material was not inexpensive. However, at the time when
RMU study was conducted, there were other cases where the applications successfully
made thousands of parts including 10 K pill dispensers through a collaboration Pichot
and Axis in Europe [9]. In another project, Axis, a service bureau located in France, was
able to make more than 1,000 parts in both ABS and Polyoxymethylene (POM). They
credited direct rapid tooling with NanoTool in cutting mold making lead times to eight
working days from four to six weeks in addition to the major cost savings obtained. Their
conclusion was that this type of direct rapid tooling application is suitable for low volume
requirements where small parts and precise tolerances are required. They also saw a great
potential for metal cladding in making NanoTool molds to be metal plated for improving
their life span.

7.2.1.2   Direct Rapid Tooling of Sand-Casting Molds

Sand casting is a metal casting process where the molten metal is poured into porous sand
molds (Fig. 7.9) where it solidifies taking the shape of the cavity. As illustrated in the
figure, a simple sand-casting setup includes *cope* (top) and *drag* (bottom) halves held by

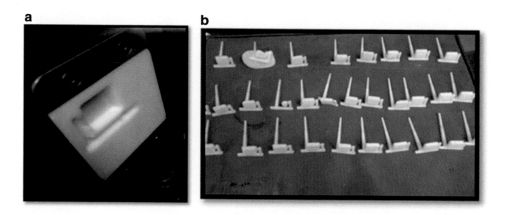

Fig. 7.8 **a** SLA printed injection molding insert and **b** resulting PC parts [5]

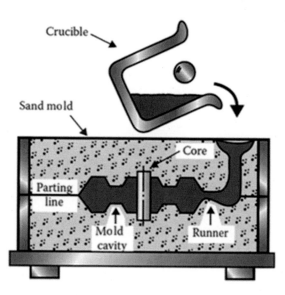

Fig. 7.9 Sand-casting process in a two-piece mold. (The riser was not portrayed.) [10]

flasks while defining the mold"'s cavity and its feeding system. *Cores* are used in making holes in these castings and a more complex three-piece mold setup includes a middle molding component, *cheek*. It also includes *risers* to serve as additional metal reservoirs in obtaining shrinkage free castings. Copes, drags, cheeks, and cores can be 3D printed directly, utilizing the Binder-Jetting method where a binder like Furan (furfuryl-alcohol based along with acid activators) is applied to the layers of green sand for making high strength cores and molds. Following table summarizes different type binders available (by the ExOne company) for making sand molds. Sand molds made by direct rapid tooling are

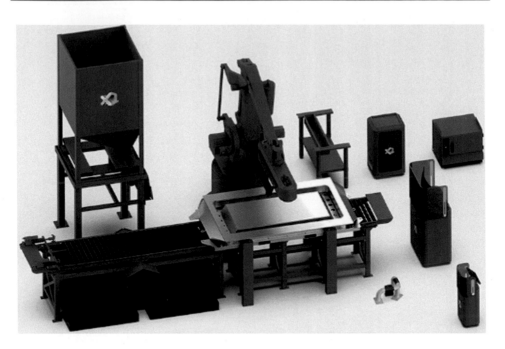

Fig. 7.10 The new ExOne system, S-Max® Flex, incorporates articulated robotics into the Binder-Jetting process with an expanded work volume [12]

slightly different from their manually prepared counterparts employed in making castings. Manual applications for ferrous castings almost always use green sand as the molding material. High-quality silica sand makes up most of the green sand, with about 10% bentonite clay as the binder, 2—5% water, about 5% sea coal to improve casting finish [11].

ExOne, a US company, has been a leader in sand mold printing with its S-series 3D printers. Its printers can be used in prototyping, serial production, and parts on demand. Its largest printer, S-Max® Flex (Fig. 7.10), has a build volume of 1900 mm × 1000 mm × 1000 mm (L × W × H = 74.8 in × 39.4 mm × 39.4 mm) with a maximum build rate of 115 l/hour (7017.73 in^3/hour) and a layer height range of 0.28—0.50 mm (280—500 μm) while its smallest printer, S-Print®, has a build volume of 800 mm × 500 mm × 400 mm (L × W × H = 31.5 in × 19.7 in × 15.4 in) with a maximum build rate of 40 l/hour (2440.95 in^3/hour) and a layer height range of 0.20—0.50 mm (200—500 μm) [12]. Table 7.3 below presents available binder types and a resulting sand mold print of this process and its outcome (an aluminum casting) are given in Fig. 7.11. The largest system of the S-series, S-Max® Flex, is its newest and incorporates an articulated robot arm to the Binder-Jetting process (Fig. 7.10) with a price tag of $1 M+?while other S-series printers being more affordable [13]. On the contrary, Voxeljet, a German company, markets VX4000 sand mold printers employing standard foundry materials. VX4000 is the largest 3D printing system for its kind in the world, with a footprint of $4 \times 2 \times 1$ m. It has a print resolution of 200 dpi, but also comes with a high price tag [14]. Thus, manufacturers, researchers, and hobbyists have been developing lower cost sand mold printers including the lead author's effort at RMU, Fig. 7.12.

7.2.1.3 Indirect Rapid Tooling of Room Temperature Vulcanization (RTV) Molds

Indirect rapid tooling can be realized by utilization of various processes including Room Temperature Vulcanization (RTV) of silicon rubber molding. Employing a metal composite pattern as a master pattern, the lead author fabricated tin-cured silicone rubber molds to cast polyurethane replicas of the pattern. The pattern was built in ProMetal"s R2 machine using a stainless steel (S4) base with bronze infiltration, as seen in Fig. 7.13a. Two other alternative methods, FDM and SLA, were also considered for making the pattern, but not chosen. A two-piece silicone rubber mold was built around the pattern with Smooth-On"s Mold Max™™ 40 (Shore A scale) [5]. Approximately 20 h was deemed suitable for curing time to obtain a solid mold. However, the molding was held longer to ensure complete curing. Mold Max™™ 40 is ideal for transferring the details of the RP pattern. It also carried the characteristics of other advanced silicone molding materials:

Table 7.3 ExOne binders available for mold making [15]

Binder type	Chemistry	Features
Furan	Furfuryl-alcohol based/acid activators	Suitable for making high strength cores and molds in various silica sand and ceramic media like cerabeads with low Acid Demand Values (ADV)
CHP	Ester-cured alkaline phenolic resole binder	Requires additional post-curing at elevated temperatures Best suited for specific mold geometries requiring maximum sand removal ability Also compatible with various silica sand and ceramic media like cerabeads
HHP	Acid-cured phenolic resole binder	Ideal for producing high hot strength dimensionally-stable cores using various silica and ceramic media like rough and fine cerabeads
Inorganic	Water-based alkali-silicate	Requires additional post-curing at elevated temperatures Additional proprietary powder and fluid additives are incorporated to optimize printed core and mold strengths as well as high-temperature casting properties

Fig. 7.11 Binder jetting printed sand molds and resulting aluminum casting [5]

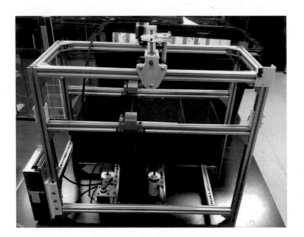

Fig. 7.12 RMU sand printer project employs the Binder-Jetting method

- Good release properties and high elongation for easy removal of complex parts,
- Low shrinkage and good dimensional stability,
- Good tear resistance and consequent mold life for casting polyurethane,
- Medium mixed viscosity and medium hardness.

The material selected for the casting process was the fast setting Smooth Cast 320, which is a polyurcthane material, also from Smooth-On. It sets in about 10—15 min and works well with the selected mold material [5]. Mann Technologies 200 Easy Release Agent was also employed in the process. The mold halves and a polyurethane replica produced by them are presented in Fig. 7.13b. Total time of 10 h was estimated to be casting lead time based on the casting cycle time of 12 min per copy, and a batch of 50 copies. When this was added to the Binder-Jetting cycle time of 53 h and mold-making

a   b

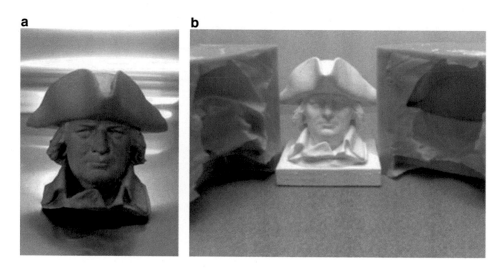

Fig. 7.13 a Master pattern **b** Two-piece silicone rubber molding and resulting polyurethane copy [16]

time of 24 h, the total lead time became 83 h. This may be longer than directly printing these copies in the SLA and FDM 3D printers, it cost only $44 per copy much less than the cost of the copies made by those printers at the time. This cost figure could have been reduced to $27 per copy if the SLA model is chosen as the master pattern. The combination of Binder Jetting/RTV molding process also yielded acceptable dimensional accuracy.

In Practice

Make a low-melting pewter copy of a large coin or a small metal object of your choice, following the steps of the indirect rapid tooling methodology below and safety precautions listed by the manufacturer including wearing personal protective equipment (PPE):

- Build a two-piece mold around your object by placing/taping your object (with a flat bottom surface) in a paper coffee cup, mix the parts A and B of the Mold Max^{TM}TM 60 material (100A/3B by weight), and pour the mix around the pattern (your object).
- After a cure time of 24 h and full curing, you can split the mold halves carefully by using an x-acto (exacto) knife (making the parting planes). Make sure that the cut for the split is clean and straight, and there is a pouring hole between the halves.

- After removing the pattern from the cavity, assemble the mold halves by using tight rubber bands, you can also use duct tape for preventing leakage at the parting line.
- Apply talc powder into the cavity, without having an excess amount, to act as a release agent.
- Melt the appropriate amount of pewter material in a safe container, and pour it into the cavity.
- Once the pewter is solidified, remove the rubber bands and demold the casting. After that point, you can post process the casting for a better look by carefully sanding its high surfaces.

Hint: You can also make a single-piece mold as seen in the tutorial from Smooth-On website [7].

7.2.1.4 Indirect Rapid Tooling of Silicone Rubber Shell Molds

This subsection focuses on a case study for fabricating shell molds for making skin for an animatronic robot at RMU. The process consisted of:

- 3D printing sections of a Robert Morris bust using an Stratasy's' Dimension (FDM) machine and joining the printed ABS sections with acetone
- Coating the joined model
- Manually modifying the joined model with help from a special effects artist to reflect aging changes over the scanned model
- Brushing platinum-cured silicone rubber molding material over the modified model
- Curing the shell mold.

The process was completed brushing additional silicone rubber into the shell mold and allowing its curing. Figure 7.14a illustrates the joined 3D printed model, shell mold made from it and the resulting animatronic robot skin is given in Fig. 7.14b.

7.3   Additive (Rapid) Manufacturing

As we know, 3D printing processes are used in making prototypes (rapid prototyping), tooling (rapid tooling), and end-use/consumer products (rapid or additive manufacturing). This section focuses on additive manufacturing of industrial parts like the burner element presented in Fig. 7.15a as well as the automotive part in Fig. 7.15b. It has two distinct folds:

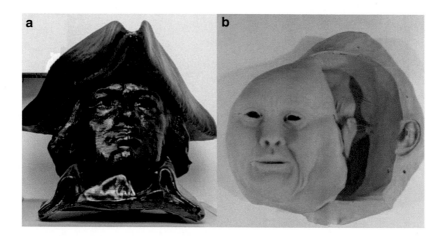

Fig. 7.14 **a** Pattern **b** Shell mold (without backing) made by brushing silicone rubber onto a modified 3D print as well as a cast silicone rubber skin for an animatronic robot skin

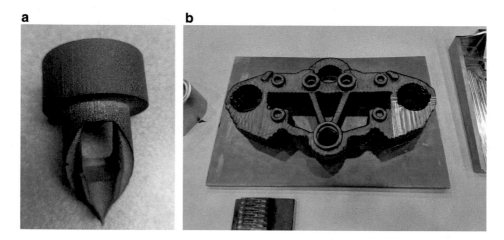

Fig. 7.15 **a** A 3D printed burner element **b** A metal AM part made by GM presented at the SME"s RAPID conference

- Job-shop efforts include fabricating of custom-made toys for children, teaching aids/tools for handicapped individuals, custom made tools for astronauts in outer space or people at remote places in case of a need, also including a war, or any custom-made solution. It may also include small volume needs for aiding industrial companies in their manufacturing operations.
  - Frazer-Nash, a UK based manufacturer, has used additive manufacturing to produce a fastener needed for its customer (Kwickbolt) with lower costs and lesser lead

times compared to traditional subtractive methods. This fastener is a temporary device which is used to align aircraft panels and fuselage during assembly [17].
- Moto2T^M race team TransFIORmers has been using metal additive manufacturing in an unconventional suspension system to gain a significant competitive advantage against its competitors [18].
• Replacing traditional manufacturing efforts in industrial companies has been a challenge requiring greater availability of process types, greater availability of material types/forms and associated characteristics, greater accuracy, greater productivity (high printing speeds), and greater capacity (larger parts and continuous processing).
- Croom Precision Medical, located in Ireland, has been working with Renishaw in validation and production of their (ISO13485 certified) medical devices. Additive manufacturing is employed in incorporating complex features of Croom"s orthopedic implants with a commercially viable cost figure [19].
- A collaboration between GE Aviation and GE additive has proved that metal additive manufacturing can compete with conventional castings on cost, resulting in 35% savings with their four 3D printed part replacements. The lead time of identifying parts for switching to 3D printing took only 10 months. The company usually spends 12–18 months for production of aerospace and land/marine turbine parts using a casting process [20].

Part consolidation is another advantage of additive manufacturing. GE Aviation was able to reduce 20-part fuel nozzle tip assembly for their LEAP engine to just 1 3D printed part, and the company's new turboprop engine assembly was simplified from 855 parts to 10 3D printed ones. The examples mentioned in this section are just a small sample of what has been happening in the additive manufacturing area. The application areas of additive manufacturing is constantly growing so are the applications in the existing areas [20].

7.4   Non-industrial Applications of 3D Printing

Non-industrial uses of 3D printing include a wide variety of areas such as food, medical, forensics, anthropology, biology, and arts, just to name a few, and also include many custom works.

7.4.1 Preserving Historical Artifacts

The process of replicating historical artifacts is not new. With the development of 2D computer scanning technologies, historians, librarians, archivists, museum curators, and amateur enthusiasts have been digitizing historical works such as books, paintings, records, documents, and making them available to the public [21]. The technology has greatly advanced from these 2-D computer scanners to 3-D digitizers. This technological advancement has broadened the users to medical technologists, anthropologists, paleontologists, primatologists, and forensic scientists. In this context, virtual reconstruction for forensic applications has been one of the growing applications of reverse engineering, replacing the hard work of skull reconstructionists [22]. The virtual reconstruction process starts with digitization of physical elements such as bone fragments of a primate or crime victim. These digitized elements are manipulated by eliminating noise, filling in missing geometric data, and assembling them within the CAD environment. The next step beyond the virtual reconstruction is to realize the CAD model via 3D printing. One of the other growing applications of the virtual reconstruction is the generation of custom implants or scaffolds for replacement of missing sections of human bones [23] which has relied on other manufacturing methods before 3D printing. The case study presented in this subsection focuses on digitizing and duplicating a pattern used in fabrication of Robert Morris, a Revolutionary War Hero who helped finance the war, statues and busts for museums and parks [24]. The pattern, shown in Fig. 7.16a was restored at the Carnegie Museum of Art in Pittsburgh, Pennsylvania. In the summer of 2007, the lead author was given the task of digitizing and duplicating the pattern without causing any damage to it. Because it was made from plaster and almost 100 years old, it had to be handled very carefully and could not be used as a molding master pattern. The pattern was scanned prior to restoration with the intention of rescanning after the completion of restoration.

A Minolta Vivid 910 scanner and Geomagic Studio software were used for the digitization (Fig. 7.16b, c) and manipulation of the digital model. The camera could scan large

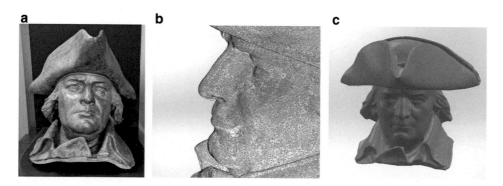

Fig. 7.16 **a** Robert Morris plaster pattern **b** Polygon file **c** STL file [16]

Fig. 7.17 Custom made
scanning platform [16]

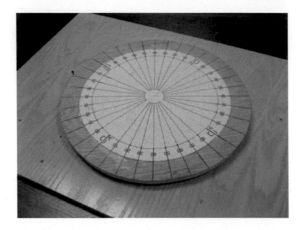

free-form objects with a dimensional accuracy of 0.1270 mm. To assist in the scanning
process, a Parker Automation 200 RT Series motor-driven rotary table with a diameter of 203.20 mm and a maximum load capacity of approximately 68 kg was available,
but not used. Because of the geometric complexity of the bust, special attention had to
be paid to cavities and shiny surfaces. Since the scanner did not have the flexibility to
reach hard-to-access details, the scanning process became more tedious than originally
expected. The main difficulty encountered during the scanning process was the special
care requirement in handling a historical artifact with a value of more than $100,000. The
pattern would fit in a 0.9144 m × 0.9144 m × 0.9144 m work envelope and weighed
approximately 27.22 kg. Such a large object with a vulnerable structure due to its fragile,
aged body required that a special scanning platform be fabricated (Fig. 7.17) due to the
small footprint of the original turntable [25].

Once placed on the platform, the object would not be moved. Since the original platform spins automatically to enable data capture through 360° during the scanning process,
it was necessary that the manual platform rotate as well. The project team calibrated the
new rotary table as if it was the one connected to the PC with the Geomagic Studio
software and accomplished each shot by matching the angle of rotation at the software
tool and the manually driven table. Various rotation angles were originally tested. After a
brief study, a rotation interval of 30° was selected as the stepping angle for the consecutive scans. As the Geomagic Studio software was instructed to rotate the original rotary
table 30° for the next scan, the investigators manually moved the second table with the
actual piece 30°. The captured data were processed within the Geomagic Studio reverse
engineering software. 122 scans were done and the resulting file was approximately 2 GB.
After manipulation of data via point cloud, polygon, and shape phases, a watertight STL

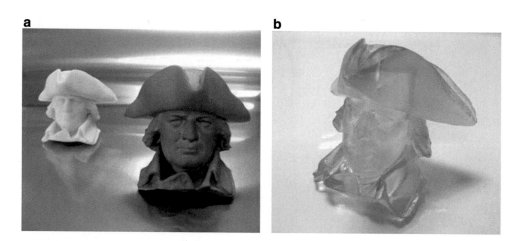

Fig. 7.18 **a** FDM (ABS) and binder-jetting (stainless steel/bronze) prints **b** SLA print ((DSM) Somos Watershed 11,120) [16]

was generated (Fig. 7.16c). This file was 200 MB in size, thus manageable by an average PC at the time and consequently 3D printable. The STL file was later used to create 3-D physical replicas of the original piece. Replicas were produced by using three machines: Stratasys Dimension Elite FDM machine, ExOne/Prometal R2 machine, and 3D Systems Viper SLA machine (Fig. 7.18a, b). Prometal print became a master pattern for making physical replicas of the bust via indirect rapid tooling as explained earlier in this chapter.

7.4.2 Food Printing

There have been a variety of food printing projects focusing on printing including possible logical applications of pizza (with its tasty layers), pasta, pan cakes, and chocolate. Food printing is still in infancy but soon households may see their first food printers due to the rapid developments. All3DP.com offers an updated and comprehensive review of this area [26]. The review also includes pros and cons of the area, commercially available machines as well as future predictions [26]:

- Today's food printers are currently aimed for gourmet dining, available in molecular kitchens and fancy bakeries, not scalable for large markets.

- They are categorized as professional printers for food enterprises (byFlow/Focus, Foodini, MMuse etc.), printers for consumers and prosumers (Mycusini, Procusini etc.), and other equipment like extruders (Cakewalk 3D, Zmorph Fab Thick-Paste Extruder etc.)
- Food printers are more appropriate for printing intricate/customized food shapes even though they can print any shapes.
- The type of food items that can be 3D printed are driven by the printing processes. FDM is the most common process allowing use of paste-like/semi-liquid materials like purées, mousses, and other viscous foods. Binder-jetting and SLS have also been explored with use of powder-based materials. These consumable materials may be cooked after the print is removed from the machine or at the build plate in case of pancakes while other can be consumed without baking or cooking.
- Benefits of printed food:
  - Besides creation of aesthetically attractive food, the process can generate personalized meals and nutrition, convert unconventional food items like protein-rich insects to more edible forms for some or mimic the texture of meat in plant-based meat. It is also easily reproducible or repeatable.
- Drawbacks of printed food:
  - Usually has relatively larger preparation times compared to conventional means along with cost of the equipment and consumable being barriers. The technology present at the time this book was written also lacked scalability.

In addition to the food preparation 3D printers explained above, there have been efforts in employing 3D printing in the form of bioprinting with stem cells for growing meat tissue for eating [27]. Bioprinting of this kind may reduce the carbon footprint associated with the meat industry and help sustain astronauts in longer travel distances in outer space. It can also be considered 4D printing since the print will change over time. NASA has already partnered with Systems and Materials Research Corporation (SMRC), located in Texas, to develop both the 3D printers and their food. The company has already developed a pizza printing system that uses a combination of powdered consumable materials containing necessary nutrients [26].

In Practice
Convert a low-cost FFF 3D printer to be used as a pancake printer (similar to the RMU one shown in the figure below). Make a PowerPoint presentation documenting your effort including a video demonstrating its successful operation.

Hint: You need to disable the heat of the FFF printer extruder within its settings/firmware of the controller or within its G-code, and use a plunger with enough batter holding capacity.

7.4.3   Medical Applications

This section is dedicated to custom medical device manufacturing, specifically orthotic and prosthetic (O&P) device manufacturing and based on the lead author and research assistant"s prior work [28]. The role of 3D and medical scanning technologies, numerical analysis, and 3D printing for additive manufacturing are covered in detail with help from ankle and foot orthotics (AFO). Fabrication of these devices laborious, inconsistent, slow, expensive, and unattractive when compared to today"s modern and technologically advanced processes [28]. Conventional practice involves educated guesses and trial and error, as the technicians try to fit patient molds (imprints) to templates and castings manually, often leading to erroneous results to be corrected for functionality and patient comfort [29, 30]. However, with the developments in 3D scanning and 3D printing technologies, the custom manufacturing of O&P devices is becoming more accurate, repeatable, and affordable, producing improved devices meeting patient"s needs, while also reducing the lead-times of new device development or manufacturing [28]. In a recent work, Constantinos Mavroidis and his collaborators conducted a test employing a motion capture system in order to compare the biomechanics of the gaits of patients wearing 3D printed

AFOs and those wearing traditionally fabricated devices. After studying the parameters, kinematics, and kinetics of the patient's' lower limb joints in the sagittal plane, they were able to determine that AFOs generated through 3D printing not only provided a suitable fit of a patient"s anatomy, but also delivered comparable functionality in to those traditionally produced [29]. Materials selection in this field is also important including use of newly developed polymer/carbon or glass fiber composites and associated processes like continuous fiber fabrication (CFF), even though the work presented in the case study simply used ABS material. SOLS uses NASA grade nylon to print their customized orthotic designs [31]. One can print an AFO with a material that is strong and highly flexible while resistant to moisture. CRP Group also notes that their material selection combines elasticity, ductility, and resistance to loading, providing substantial applications to devices that require strength and support while having the ability to mimic the movements of the foot and ankle through a normal gait cycle [32].

The automated custom O&P device development and manufacturing process starts with acquiring the necessary information from patients, whether it is based on external (laser or light-based 3D) scanners or internal medical imaging tools such as computerized tomography devices (CAT scanners), magnetic resonance imagers (MRIs) or ultrasonic devices. The patient"s information is then manipulated by removing the noise, patching its voids, and smoothened to be converted into a computer-aided design (CAD) model. The resulting CAD model is employed in the design and development of the custom device by selecting the materials and automated manufacturing processes as well as completing the detail design of the device. This process may also include prototyping via 3D printing or CNC machining before an actual or test device is manufactured. After the fabrication of the actual or test device, it is finished by applying post-processing operations and fitting them to the patients for actual use or testing for further improvements in the individual device.

The traditional method of making custom fit ankle and foot orthotics (AFO) may take a lead-time of approximately 4 h and follow a labor intensive sequence that includes making markings on a sock that is worn by the patient for helping technicians where to apply modifications when necessary, wrapping the leg with plaster to cast it, filling the cast to generate a bust (pattern), embedding staples into the bust for surface markers, vacuum forming thermoplastic material to fabricate the AFO around the bust, and removing unwanted plastic from the AFO. Voids resulting from cutting the mold are filled for future use of the bust as the busts are stored for a few months [33]. The following subsection sketches the efforts of developing an automated method based on 3D scanning and printing as an alternative to the traditional method mentioned above.

This section presents the work done by the supervision of the lead author at RMU [28]. Krivoniak took two different approaches to develop an automated method to manufacture a customized AFO. Both followed a sequence involving 3D scanning of the lower leg anatomy (using an anatomical model), data manipulation for preparation to design, 3D modeling, and 3D printing (Fig. 7.19). However, the 3D scanning method and details

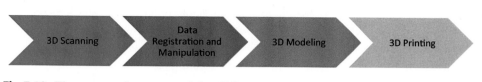

Fig. 7.19 The automated process workflow [34]

of the other sub-processes differed between the two approaches with a FARO Arm scanner or Xbox Kinect sensor). Even though the methodology described below focused on developing and streamlining an automated process for making custom AFOs, it can still be used as a basis for other research efforts:

- FARO Arm data were fed into the Geomagic Studio software to generate a point cloud replica of the lower leg and foot of the manikin. Multiple scans were slowly taken from various angles, being certain to capture all necessary regions of the anatomy for fabricating a custom fit AFO. Due to the scanner arm being attached to a benchtop (Fig. 7.20a), the range of the person performing the scanning was restricted, making it critical that every area of the lower leg and foot was captured with extreme focus and attention to detail. The scans were then manually and globally registered and merged to create a watertight model. This model then became an accurate, fine-tuned model (Fig. 7.20b) of the anatomy following the future steps presented below [34].

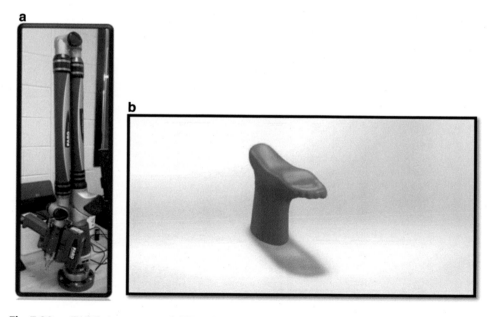

Fig. 7.20 a FARO Arm scanner **b** Watertight scan model [28]

Data manipulation phase of the project involves point, polygon, and shape sub-phases. Geomagic Studio software was used to process the data acquired from the FARO Arm. It also helped to reduce noise including the outlying points while retaining the scan"s detail. Following information summarizes the second phase:

- The "Points Wrap" command helped reduce the points and generate a uniform.wrap file.
- The polygon phase capabilities of Geomagic Studio were used to repair intersections, fill holes, and refine floating data and edges. The "'Select by Curvature" command was employed to relax the structure while retaining detail.
- A non-uniform rational b-spline (NURB) surface was generated to finalize the mesh structure and export it as an (Stereolithography).STL file.
- Once a 3D mesh has been generated, the number of triangles comprising the mesh must be reduced in order to lower computing lag time. MeshLab software was utilized to reduce the number of triangles to 7000 as shown in Fig. 7.21. Gradual steps were used to reduce the number of triangular facets. Each step reduced the number of faces by half of the original number. After each reduction, the mesh was visually inspected to ensure that no structural deformations to the lower leg and foot had occurred as a result of the mesh regeneration.

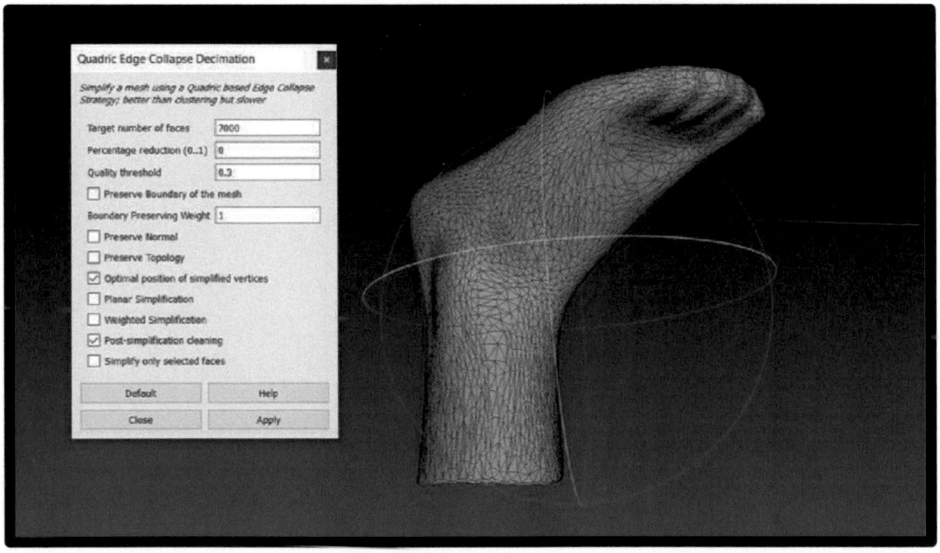

Fig. 7.21 Reduced mesh structure with 7000 triangles [28]

- An Xbox 360 Kinect was used to capture the lower leg anatomy in the second approach. The Kinect"s human recognition capabilities stem from its 2D and 3D camera systems. The Kinect transmits an infrared pattern onto the subject that is being scanned. The pattern is generated with diffraction grating. Once the pattern is transmitted and reflected from the scanned object's surface, it is captured by the Kinect"s camera. The Kinect then reads and stores these measurements in the Cartesian coordinate system including the distance and deviation of the data to the zero axis [35]. To capture the scan, the SDK Browser that correlates with Kinect"s Windows compatibility and the Skanect software package needed to be downloaded and available on a nearby PC. The Kinect was plugged into the power, as well as the nearby computer through a USB cable attachment. The Kinect"s wire attachments did not pose many restrictions. The wires were of ample length and allowed for the person performing the scanning to move freely around the subject being scanned. After one full revolution of the Kinect scanner around the lower leg and foot, the scanned data appeared in the Skanect window and was ready for the data manipulation phase of the process [34].
  - Skanect software was used to reduce the mesh structure to 10,000 polygons. The software"s features were also utilized for removal of outlying points, smoothing, and the generation of a watertight.STL file. This file could be imported into the Meshmixer program and the previously listed steps could be taken to generate an AFO from this scan as well [28].

Following describes the design of the AFO, after both approaches yielded their own.STL files. The Meshmixer program allowed simple, yet detailed modifications to be made to the.STL file of the scan mesh geometry. Thus, with the Meshmixer, the AFO could be produced directly from the patient"s original scan without the concern of losing much detail or accuracy provided by the scan. After the.STL file was imported into the Meshmixer software, the "offset" command was used to set the contact surface of the AFO 3 mm from the surface of the lower leg and foot, allowing for patient comfort and a realistic fit. Then, the "extrude" command was used to produce depth for the reality of an AFO and for the 3D printing process. Next, the scanned structure was modified by removing excess material and utilizing "'smoothing" and "'sculpting" commands to produce a clean and detailed structure of an AFO as shown in Fig. 7.22 [28].

A Stratasys uPrint SE (Fig. 7.23a), an FDM machine, with a build volume of 203 × 152 × 152 mm (8 × 6 × 6 in), was used to print the AFO model in scale for form checking and design evaluation purposes. The uPrint SE employed ABS material with a layer thickness of 0.254 mm (0.010 in) to print the non-functional prototype (Fig. 7.23b) [28].

This section compares the two approaches taken for designing and manufacturing custom AFOs and discusses possible future work in this area. Table 7.4 presents the lead times of the scanning, data manipulation, and modeling phases for both approaches. The total lead-time for the three phases of the FARO Arm option resulted in 1 h and 45 min, while the XBOX 360 Kinect phases were completed in less than an hour (50 min). The

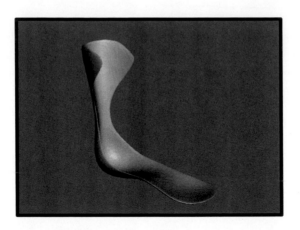

Fig. 7.22 Customized AFO design as a result of 3D scanning, data registration/manipulation and modeling [28]

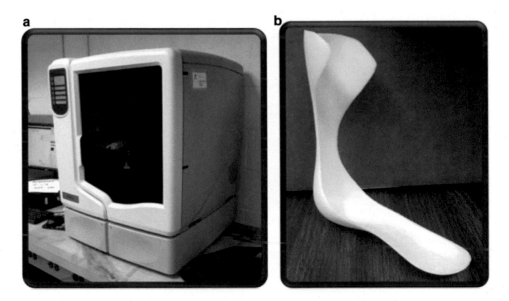

Fig. 7.23 **a** uPrint SE printer and **b** the 3D printed AFO prototype [34]

times listed above are approximations based on the average time to complete each process after several trials were performed. It is important to note that these times can vary slightly depending on the processing speed of the computers used, as well as the user"s skill level regarding scanning, data manipulation, and modeling. It is also worth noting that the times to model the AFO that are listed above are the same for each of the scanning methods because the same process was used in both cases. The time to 3D print the

Table 7.4 Time to complete design process [28, 34]

Scanning method	Time to complete scan (min)	Time to register/manipulate data (min)	Time to model AFO (min)	Total Lead-time to obtain 3D print ready models (min)
FARO arm laser scanner	25	40	40	105
Xbox 360 kinect	2	8	40	50

AFO model was not included since a scaled prototype of the device was printed in this study [28].

Additional comparisons of the two approaches are summarized in the following table, Table 7.5.

The question imposed by this study inquired if it was possible to devise an efficient and effective method to manufacture a customized AFO that is fast, consistent, repeatable, comfortable for the patient, and flexible to design modifications. The factors of speed, consistency, repeatability, comfort, and flexibility have been addressed in previous sections.

However, it is also important to note the trade-off of high-accuracy and low-cost associated with 3D printing to traditional methods. At the time this project was completed the authors felt that the highly accurate method of using 3D scanning to reproduce a patient"s exact anatomy traded with the one-time investment in the hardware and software needed to execute the process. Since then portable 3D scanners such as Structure Sensor (Fig. 7.24) have taken the role of a trial device like Kinect with only an investment of a few hundred dollars. The further advancements of 3D printing have also led

Table 7.5 Comparison of two approaches [34]

Scanning method	Scanning process	Repeatability of process	Flexibility of process	Consistency of process and deliverable
FARO arm laser scanner	Mounted to a benchtop, the user had little freedom around the object being scanned	Both methods proved to be reliable in terms of being a repeatable method for producing custom AFOs	Three phases of the both methods proved to be flexible for allowing changes for patient's specific diagnosis quickly	Both methods successfully eliminated the human interference and associated human errors
Xbox 360 kinect	Device office its user a greater range of motion for scanning, even with presence of long cables			

Fig. 7.24 Structure sensor attached to an iPAD being utilized in a scan

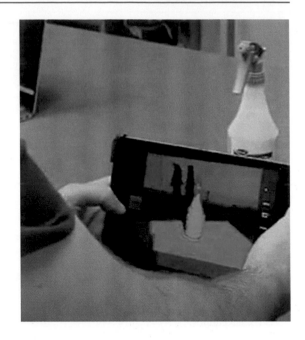

to development of new materials including exotic composites, reduction in printing costs along with reduced waste and shortened development and manufacturing lead times, and removal of labor, its cost and associated human error. Finally, this technologically superior method can aid clinicians in their ability to produce more durable and aesthetically pleasing devices, in turn, attracting a greater market of consumers [28, 34].

Review Questions

1. What are the three industrial uses of 3D printing?
2. Define the term ""rapid prototyping"".
3. What is virtual prototyping and what is its role in product or machine design? Elaborate briefly.
4. What are the four ways of rapid prototyping? Elaborate on each one briefly.
5. List the advantages of 3D printing over CNC machining when used for rapid prototyping or in general.
6. What are the 3 F's of prototyping? Explain each one with an example.
7. Define the team ""rapid tooling"".
8. What is the difference between direct and indirect rapid tooling? (Hint: Make sure to include their uses in your comparison, not just the nature of them).
9. Review the direct rapid tooling cases given in the chapter including the manufacturing processes associated with them.

10. Review the indirect rapid tooling cases given in the chapter including the manufacturing processes associated with them.
11. Define the term ""rapid (additive) manufacturing"".
12. Make a mindmap for the main points given in the food 3D printing subsection and try to enhance it.
13. What is the process workflow of digitizing a historical artifact? Explain each step briefly?
14. What are the advantages of automated AFO fabrication compared to the conventional methods? Elaborate each item briefly.
15. Review the custom AFO fabrication workflow. Is it similar to the custom-made bone replacements?

Research Questions

1. Conduct a literature review to find a case study on alpha and beta prototyping stages in product development and summarize it in two paragraphs total.
2. Conduct a literature review to find a case study on additive manufacturing where a sizable product volume is desired. Summarize the case in a PowerPoint slide.
3. What are properties required from materials like Mold Max^{TM}™ 60 for casting low-melting pewter?
4. What are the standards applicable to food printers? Conduct a literature review and summarize your findings in at least two PowerPoint slides.
5. Learn how to scan a physical object using a Kinect device and 3D print the digitized geometry.
6. Conduct a literature review on the ethics issues in bioprinting including making meat from animal stem cells and write a one-page summary statement.
7. What are the device classifications/engineering standards associated with the medical devices mentioned in this chapter? Make sure to address each one.

Discussion Questions

1. Learn about the Ultimaker"s S5 Pro Bundle. Let"s assume, this machine is being considered for making a large volume of 3D printed light diffusers weekly. What will be the advantages/disadvantages of such a set-up?
2. Can the system mentioned in the previous question be (further) automated?
3. Can 3D scanners be used in digitizing art paintings? Elaborate briefly.
4. Expand the benefits and drawbacks of food 3D printing discussion presented in the chapter. Use examples for justifying your points.
5. Discuss the future of food 3D printing and prepare a PowerPoint presentation with no more than 5 slides.

References

1. Ulrich, K.T., Eppinger, S.D. (2012). *Product Design and Development 5th Edition.* New York, NY: McGraw-Hill Irwin.
2. Joo, W., Carlsen, R., Sirinterlikci, A., McChesney, C., Dodds, N., Proctor, B., Leimkuehler, P., Leimkuehler-Mullin, L. (2020), Adopting Additive Manufacturing Technologies for Orthotics and Prosthetics. Unpublished Robert Morris University Research and Grants Expo Poster.
3. DeGrosky, J., Workmaster, C., Dodds, N., McChesney, C., Minarik, N., Proctor, B., Leimkuehler-Mullin, L., Leimkuehler, P., Sirinterlikci, A., Carlsen, R., Joo, W. (2022). Adopting Additive Manufacturing Technologies for Orthotics and Prosthetics, Unpublished Robert Morris University Research and Grants Expo Poster.
4. Two Types of Rapid Tooling for Prototyping, Design Blog, located at: https://www.pacific-research.com/two-types-of-rapid-tooling-for-prototyping-prl/, accessed September 20, 2022.
5. Sirinterlikci, A., Czajkiewicz, Z., Doswell, J., Behanna, N. (2009). "Direct and Indirect Rapid Tooling"", 2009 Rapid/3D Scanning Conference, Chicago, IL.
6. Wimpenny, D.I., Bryden, B., Pashby, I.R. (2003). Rapid Laminated Tooling, Journal of Materials Processing Technology, 138 (1-3), pp. 214—218.
7. How to Cast Pewter into Mold Max® 60, located at: https://www.smooth-on.com/tutorials/casting-pewter-mold-max-60/, accessed September 20, 2022.
8. Mengel, M. (2005). Lessons in the Use of Electroformed Nickel Tools, Solutions to problems mold makers face with today"s styling requirements in the instrument panel tooling industry, located at: https://www.moldmakingtechnology.com/articles/lessons-in-the-use-of-electroformed-nickel-tools, accessed September 20, 2022.
9. New RP Technology Reduces Costs, Lead times (published on 08/01/2008), located at: https://moldmakingtechnology.com, accessed September 20, 2022.
10. Sirinterlikci, A. (2013). Manufacturing Processes and Systems Chapter, *Industrial and Systems Engineering Handbook* 2nd *Edition.* Boca Raton, FL: CRC Press.
11. Foundry sand, Material Description—FHWA-RD-97–148, located at: https://www.fhwa.dot.gov/structures, accessed September 20, 2022.
12. 3D Printing Systems, Sand 3D printers, located at: https://www.exone.com/en-US/3D-printing-systems/sand-3d-printers, accessed September 20, 2022.
13. ExOne S-Max Review, an Industrial 3D Printer (Binder Jetting)–Aniwaa, located at: https://www.aniwa.com/product, accessed September 20, 2022.
14. Industrial 3D Printer, Serial Production, VX4000, located at: https://www.voxeljet.com/industrial-3d-printer/serial-production/vx4000/, accessed September 20, 2022.
15. 3D Materials & Binders–ExOne, located at: https://www.exone.com/en-US/sand casting, accessed September 20, 2022.
16. Sirinterlikci, A., Uslu, O., Behanna, N., Tiryakioglu, M. (2010). Preserving Historical Artifacts through Digitization and Indirect Rapid Tooling, International Journal of Modern Engineering, 10 (2), pp. 42—48.
17. Metal additive manufactured parts for aircraft assembly, located at: https://www.renishaw.com/en/metal-additive-manufactured-parts for aircraft assembly-44235, accessed September 20, 2022.
18. Metal 3D printing pushes the boundaries in Moto2TM through defiant innovation, located at: https://www.renishaw .com/en/automotive--44456, accessed September 20, 2022.
19. Additive Manufacturing for serial production of orthopedic implants, located at: https://www.renishaw.com/en/medical-and-healthcare--44309, accessed September 20, 2022.

20. Game Change: Four parts proving additive manufacturing can compete with casting on cost, located at: Blog.geaviation.com, accessed September 20, 2022.
21. Rosenzweig, R. (2005). Digital History: A Guide to Gathering, Preserving, and Presenting the Past on the Web. Philadelphia, PA: University of Pennsylvania Press.
22. American Academy of Forensics Sciences, Resources, located at: http://www.aafs.org/default. asp?section_id=resources &page_id=choosing_a_career, accessed December 1, 2008.
23. Zollikofer, D. (2005). Virtual Reconstruction: A Primer in Computer-Assisted Paleontology and Biomedicine. Hoboken, NJ: John Wiley and Sons.
24. Robert Morris University, History and Heritage, located at: http://www.rmu.edu/public-rel ations-andmarketing/content/history-and heritage.aspx?it=&ivisitor= , accessed December 1, 2008.
25. Sirinterlikci, A., Uslu, O., Czajkiewicz, Z. (2008), Replicating Historical Artifacts: Robert Morris Case Study, SME 3D Scanning Conference, Lake Buena Vista, FL.
26. 3D Printed Food, located at: https://all3dp.com/2/3d-printed-food-3d-printing-food/, accessed September 20, 2022.
27. Scientists create First 3-D Printed Wagyu Beef–replicate cut"s specific arrangement of muscle, fat, and blood vessels, located at: https://www.smitsonianmagazine.com/, accessed September 20, 2022.
28. Krivoniak, A., Sirinterlikci, A. (2017), 3D Printed Custom Orthotic Device Development: A Student-driven Project, American Society for Engineering Education (ASEE) Conference, Columbus, OH.
29. Mavroidis, C., Ranky, R. G., Sivak, M. L., Patritti, B. L., DiPisa, J., Caddle, A., Bonato, P. (2011). Patient specific ankle-foot orthoses using rapid prototyping. Journal of Neuroengineering and Rehabilitation, 8(1), pp. 1—11.
30. New 3D printing technology helps enable customized, in-office-printed orthotics, Los Angeles: Anthem Media Group. Retrieved from, located at: https://reddog.rmu.edu/login?url=http://sea rch.proquest.com/docview/1718199313, accessed April 12, 2016.
31. SOLS; SOLS and WebPT partner to bring custom 3D foot orthotics to rehab therapy professionals nationwide (2015). Journal of Engineering, retrieved from https://reddog.rmu.edu/login?url= http://search.proquest.com/docview/1654415561, accessed April 12, 2016.
32. 3D-Printed orthotics from reinforced laser sintering materials (2015). Medical Design *Technology,* retrieved from https://reddog.rmu.edu/login?url=http://search.proquest.com/docview/168 7713503, accessed April 12, 2016.
33. How Prosthetic Sockets Are Made, located at: https://risprosthetics.com/how-prosthetics-soc kets-are-made/, accessed September 20, 2022.
34. Sirinterlikci, A., Krivoniak, A. (2017), "Development of a Custom Ankle-Foot Orthotic Device"", 2017 Rapid +TCT Conference, Pittsburgh, PA.
35. Zug, S., Penzlin, F., Dietrich, A., Nguyen, T. T., and Sven, A. (2012). Are laser scanners replaceable by Kinect sensors in robotic applications? Institute of Electrical and Electronics Engineers. Retrieved from https://pdfs.semanticscholar.org/00a7/bb020b5f1eea7311439289f8 c15f79c62429.pdf, Accessed April 12, 2016.

Computer-Aided Engineering (CAE) and Industrial Internet of Things (IIoT)

8

8.1 Introduction to IoT/IIoT

Engineers of today in all disciplines can no longer be bound by their own field's knowledge and skills [1]. Mechanical or manufacturing engineers working in the automotive field need to understand some fundamental electrical, electronics and computers engineering, computer science considering that today's cars may have a few hundred to a few thousand chips (integrated circuits (ICs)), in addition to having an electrical system. Another requirement may be understanding the sensors and data acquisition tools used in manufacturing processes and systems including those for assembly, and tracking of parts or products (*asset management*) throughout the supply-chain. Internet of Things (IoT) has emerged as a major technology in providing measurements, monitoring, control, and analysis [1] in large scale systems including industrial and manufacturing (Industrial Internet of Things—IIoT). IoT integrates components of these systems and makes big data and associated knowledge base available for artificial intelligence (AI) and machine learning (ML) efforts in improving them. A wide range of needs, anywhere from a house light controlled by a phone app, to safety sensors in a factory being monitored remotely, to predictive maintenance of an industrial machine, to improving the performance of overall systems by increasing uptime and efficiency, can be satisfied with IoT.

Successful examples of IoT and IIoT in Industry 4.0 efforts have been increasing steadily in the last few years, and encompass construction and building automation, urban planning, and transportation systems in addition to the areas mentioned above [1]. In one case, the lead author was impressed by Neptun Glass srl, located in Rovello Porro, Italy, while visiting the company in November 2018. Neptun srl is the maker of glass processing machinery including vertical working centers, vertical and horizontal washing machines, straight line edgers along with providing robotic solutions in glass handling [2]. During the visit, the lead author and his students were exposed to the Neptun machines' black boxes for the IoT functions as well as their data acquisition process via their cloud-based interface including the performance of some of their machines.

Understanding IoT depends on understanding the following factors [1]:

- IoT origins and business perspectives that drive the need for IoT/IIoT.
- IIoT infrastructure (architecture) including its hardware (sensing, networking, computing), and software components as well as system and data management, and their security.
- Successful IIoT applications (and possibly the unsuccessful ones) in a range of domains including manufacturing, building and construction, energy, and urban planning and management.

8.1.1   IoT Systems Architecture

IoT architecture consists of devices like sensors/switches and actuators, its network structure and cloud- or edge-based technology that allows these devices to communicate with each other [3]. The IoT process starts with sensing, followed by acquisition and transmission of the sensor data, its processing, associated analytics and decision making, and concludes with its application and management aspects [4]. IoT allows transformation of data into usable information through data analytics. Consequently, companies can act by employing automation, AI or ML to improve their operations or solve their problems. The IoT process is explained below with more details [3] (Fig. 8.1):

- *Sensing* (also called *perception*) layer includes the physical devices like sensors and actuators of all sorts such as health monitors, lighting systems, and autonomous vehicles. At this layer, sensors collect data from their environment. These devices are often inexpensive and can collect large scale data for processing. Atmospheric data such as moisture levels, air temperature, wind patterns as well as solid temperature are collected in agricultural IoT applications to improve crop yield.
- *Network* layer moves (*transports* or *transmits*) the collected data to the cloud or edge device for later processing. This layer employs local networks and Internet gateways for transmission. Companies rely on cellular and Wi-Fi networks in this layer such as Cellular 4G LTE /5G, Bluetooth, and Low-Power Wide Area Networks (WAN). Another critical factor for this layer is the transport protocol, which should allow reliable moving of the data to the closest Internet gateway.
- *Data Processing* layer transforms the data into useful information involving the cloud- or edge-server application. AI and ML is often employed in the analysis process. In a

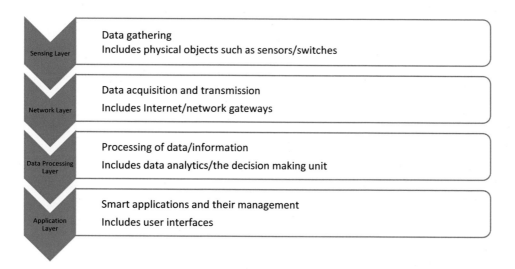

Fig. 8.1 IoT systems architecture and its process [4]

building automation example, IoT sensors allow the system to measure temperatures and humidity levels within the building, and the server can identify the issues in this layer and help correct them by talking to heating the ventilation air-conditioning system (HVAC) in the following layer.

- Even though processing of data occurs autonomously, and so does the *Application* layer, both are determined by the IoT administrators. The design of this layer includes device orchestration, creation of rule sets, and setting service level agreements by the administrators.

If we can add a fifth non-technical layer to the IoT process, it would be the *business* layer, or the intelligence needed for business decision-making. The reports and live dashboards provide the information required for the business analysts as in the case of employing IoT to control the energy costs of a condo complex [3].

8.1.2   IoT Use Cases

IoT (or IIoT) has a strong potential to impact many applications or industries, and is seen limitless by many. Following examples only present a small fraction of IoT uses [3]:

- IoT in healthcare: With improved data collection and accessibility, doctors can make better decisions while caring for their patients, including emergencies. Wearable devices like continuous glucose meters (GCMs) and smartwatches can collect vital data and share them with the primary care or other physicians involved.
- IoT in manufacturing: Allow manufacturers to collect data on their machines including the overall manufacturing system. Uses of such an effort encompass monitoring production rates/cycle times and other associated factors for meeting demands (takt time)/forecasts or accomplishing good predictive maintenance of the machines by measuring and recording their vitals.
- IoT in agriculture: Is utilized in predicting crop outputs and increasing the yield, and even possibly managing the whole growth and harvesting processes autonomously at a farm. Greater availability of 5G networks will help this type of effort, making farmers successful in their challenge. If successful, they may just use their smart devices including phones to monitor and control their farms' activities.

In Practice

Multiple educational IoT kits with a reasonable cost are available including the Arduino Oplà IoT Kit and Arduino Explore IoT Kit. More information on the Arduino Explore IoT Kit can be found at [5]. This kit includes ten step by step activities and open challenges. The authors encourage use of this or another similar kit like Oplà to get familiar with IoT [6]. The Oplà kit comes with an Arduino MKR IoT Carrier (with a color display, an IMU, touch buttons, environmental sensors, a battery charger, a buzzer, and two relays). It also includes an Arduino MKR Wi-Fi 1010 board, two plug and play cables for the sensors, a battery connection cable, a motion sensor, a moisture sensor, a plastic enclosure, and a USB cable as illustrated in the figure below. Purchasers could also subscribe to Create Maker Plan for 12 months for free, at the time when this book was being written. The kit comes with eight self- assemble projects that include:

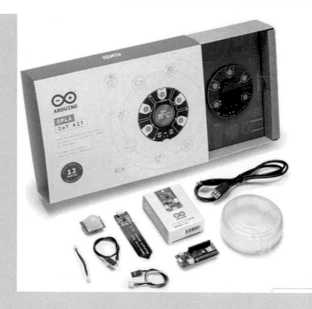

- Remote Controlled Lights for changing color, light modes and switching on/off via your mobile phone
- Personal Weather Station for recording and monitoring local weather conditions
- Home Security Alarm for detecting motions and triggering warnings
- Solar System Tracker for retrieving data from planets and moons in the Solar System
- Inventory Control for tracking goods in and out
- Smart Garden for monitoring and controlling the environment for your plants
- Thermostat Control-for smart controls in heating and cooling systems
- Thinking About You—sending messages between the Oplà and the Arduino IoT Cloud.

8.1.3 Present State and Future of IoT

World Economic Forum (WEF) promoted its efforts on accelerating action on Industry 4.0 during its Global Technology Governance Summit [7]. Its efforts highlight the impact of Industry 4.0 in different fronts including business disruption, disruption to jobs and skills, innovation and productivity, ethics and identity, inequality, agile governance, and fusing technologies (Fig. 8.2). Thus, Industry 4.0 is reshaping entrepreneurship, future of enterprises, and helping achieve a circular economy, inclusive growth framework, sustainable development while improving mental health of the involved [8]. According to the WEF, the IoT, as one of the facets of Industry 4.0, is powered by three emerging technologies: AI, 5G, and Big Data. A more connected future can be achieved *by a fusion of AI and IoT*, or basically *AIoT*. Thus, AIoT is becoming more mainstream as it transforms our lives and work [7]. The four major segments of AIoT are *wearables, smart homes, smart cities, and smart industry*. WEF's business analysis on these segments are summarized below: [7]

- Wearables: Involves smartphones, sensors like CGMs, actuators like insulin applicators worn by humans. The aim for this segment is sport and fitness as well as healthcare purposes. The tech research firm Gartner estimates this segment's revenue to be $87 billion by 2023.
- Smart Homes: Includes smart appliances, lighting, and other devices including building automation. Being responsive to homeowners' habits or improving energy efficiency are some of the benefits of this segment. Smart home market could experience *a compound annual growth rate (CAGR)* of 25% between the years of 2020–2025, to reach $246 billion.

Fig. 8.2 Industry 4.0 and its impact in different fronts [8]

- Smart Cities: The goals of making cities safer and more convenient places to live will involve AIoT as the systems work to improve public safety, transport (via *Intelligent Transport Management Systems (ITMS)*), and energy efficiency as well as sustainability aspects of them.
- Smart Industry: Involves real-time data analytics on processes, and overall systems, including their supply-chain. Systems will become more efficient and human error including the ones in-decision making processes can be averted. Gartner estimates that by 2022 over 80% of enterprise IoT projects will actually be AIoT projects.

Global newswire recently reported that a market research study done by Facts and Factors estimated the size of the global IoT market to reach $1,842 billion in 2028 after seeing a CAGR of 24.5% within the period of 2021–2028. This figure was $310 billion in 2020. Facts and Factor analysis included inputs from large corporations such as Microsoft, Amazon Web Services, IBM, Google, Cisco, HP, Intel, PTC, and Oracle [9].

Ericcson research predicts that the advances to the industrial Internet and associated network technology will be accelerated, facilitating IoT use at hyperscale. Handling of huge volumes of actionable data will be realized, leading to diversification of business processes, and opening entire new streams of revenue [10]. Thus, availability of 5G is the key to the development of IoT systems and their adoption, considering that it is approximately 100 times faster than its counterpart in 4G. 5G networks will also make IoT systems more reliable as it improves speed and data transmission [11]. However, all of this may come with its issues including costs and network data control restrictions by utility companies as well as security. Future IoT applications will have very little limits, and involve human and machine interactivity, and human to human interactions (*Human 4.0*) relying on 3D audio and haptic sensations and immersive MR applications [10]. Table 8.1 gives examples of current IoT technologies and their present and possible near future applications [12].

Table 8.1 Present and future IoT applications [12]

Technology category	Present	Tomorrow
Edge computing	• Smart thermostats • Smart appliances	• Home robots • Autonomous vehicles
Voice AI	• Smart speakers	• Natural language processing (NLP) • E-payment voice authentication
Vision AI	• Massive object detection	• Video analytics on the edge • Super 8 K resolution

8.2 CAE for Virtual Prototyping, and Digital Twins

Computer-Aided Engineering (CAE) has been serving engineers and engineering scientists for decades. In addition to digital twins, this section focuses on CAE's application to manufacturing. There are multiple numerical methods of CAE including Boundary Element Method (BEM), Finite Difference Method (FDM), Finite Element Method or Analysis (FEM/FEA), and Finite Volume Method (FVM). These methods are used in a variety of physics-based simulations or analysis. These simulations can be conducted in different situations: *in-transient/instationary form* (before the system reaches a balance after being loaded/constrained or, for a system that never reaches a balance) or in *steady-state* (after the system reaches a balance).

8.2.1 Introduction to Numerical Methods

Following are the major analysis types most engineers work on:

- *Structural analysis* aims to determine distribution of internal forces and moments, stresses, displacements and strains over a whole engineered product or structure (or a part of it) under certain loading/constraints, which can be *fixed/static* (not changing over time) or *dynamic* (time-variant) in the case of mechanical fatigue analysis. The loading types mentioned here, static or dynamic, are also applicable to the analysis types given below.
- *Thermal analysis* aims to determine the heat transfer (heat flux) and temperature distributions within a system which experiences heat being added to it or removed from it during a process. Some thermal analysis cases may involve evaporation, liquification, or solidification (*phase change*) and *mass transfer* as well. Three forms of heat transfer (conduction, convection, and radiation) as well as heat storage are found in this type of analysis.
- *Computational fluid analysis (CFD)* focuses on understanding a fluid (liquid or gas) flow case with changes in flow rates, and associated velocities and pressure, energy types and energy losses due to friction and geometric factors, as well as changes in a fluid's density. A basic CFD tool is often used for pressurized systems and may include "steady-state and instationary flows, incompressible viscous flows, compressible inviscid flows, laminar and turbulent flows, non-isothermal flows, flows with varying density and viscosity, flow through porous media, two- and multi-phase flows, fluid structure interaction, and analysis with user-defined equations and partial differential equations (PDE)" [13].
- *Electromagnetic analysis* deals with static, time-varying electromagnetic fields, and frequency-domain-based electromagnetic fields. This type of analysis involves solving

for flux, flux density, flux strength, core losses, and force and torque output of the armatures of electrical machines.

- *Biological analysis,* even though it is possible and done frequently enough, in some cases i.e. "it may deal with complex micro-level solid–fluid boundaries in the hydrated soft tissue requiring microscopically fine finite element (*mesh*) structures, making FEM unrealistic" [14].
- *Coupled analysis:* This is a combination of the analysis types presented above. In the case of a thermal/structural analysis of the die-casting dies, it can be handled as a *fully-coupled* analysis where both types of analysis conducted concurrently, and in the case sequentially-*coupled* analysis, the thermal analysis can be conducted first followed by the structural one employing the results of the thermal analysis. Thermal fatigue, creep, or electromagnetic/thermal analysis are some of the examples of coupled analyses for studying interaction of two different physical phenomena.
  - *Other:* Numerical methods may involve impact of events like chemical reaction or radiation exposure to flow [15].

The boundary element method (BEM) is a numerical computational method of solving linear partial differential equations which have been formulated as *integral equations* (*boundary integral form*) in multiple physics problems. The method is very useful where the solution domain is very large to be handled by methods like FEM. These integral equations are considered as an exact solution of the partial differential equations governing the phenomenon, unlike the other major numerical methods. This method uses the given boundary conditions as the boundary values into the integral equations. After the initial step, the integral equation can then be used again to calculate the solution at any desired point within the interior of the solution domain [16]. Figure 8.3 shows a comparison of FEM and BEM meshes, FEM mesh is applied on the 2D geometry given in the figure, whereas the BEM mesh is only applied to its boundary in the form of linear boundary elements. If the problem geometry was a 3D one, the BEM mesh would be again applied onto the boundary surfaces, along with the boundary conditions as the FEM mesh would be applied throughout the object's volume.

The finite-difference method (*FDM*) is the most direct or simplest approach to discretizing PDEs [17]. A set of discrete equations, called *finite-difference equations,* are used in the analysis *over a grid*. This method provides an efficient solution for rectangular or block-shaped geometries, by approximating derivatives with finite differences (in forward, central, and backwards schemes). Figure 8.4 is illustrating a die-casting part model saved in. STL format for an analysis in an FDM-based tool, MAGMASOFT [18]. Mathematical formulations generated by FDM are identical to those of FEM when applied to a simple geometry (resulting in the same grid). This type of analysis is used in thermo/fluids including meteorological, astrophysical, and seismological analysis [17].

The finite element method (FEM—also referred to as Finite Element Analysis (FEA)) divides the object in concern into small but finite(-sized) elements of geometrically simple shapes. The collection of these finite-elements constitutes the mesh structure [17].

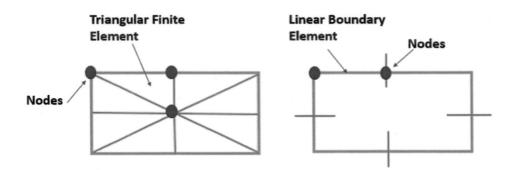

Fig. 8.3 2D Mesh structures of FEM (whole solution domain/object in concern is discretized) and BEM method (only the object's boundary is discretized)

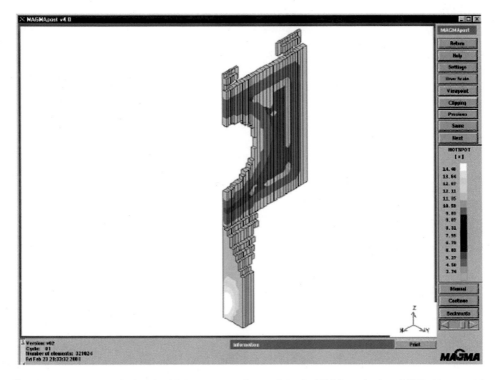

Fig. 8.4 Hot-spot analysis of a high-pressure die-casting via FDM method and .STL geometry in MAGMASOFT software [18]

According to Ansys resources [19], there are three methods for converting 3D geometries into finite-element mesh: one that uses tetrahedral elements (better fit for complex geometries), one that uses hexahedral elements (also referred to as bricks, giving more accurate

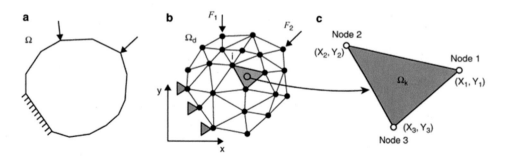

Fig. 8.5 A 2D structural finite element problem **a** geometry and problem definition **b** boundary conditions and mesh structure **c** an individual element with its nodes and DOF [20]

results at lower element counts), and a combination of both—hybrid meshing. The element types may carry different numbers of nodes for different types of analysis, linear and quadratic. A special 4-node tetrahedral element is a pyramid, and the use of triangular prisms (wedges) are also common. Once a mesh structure is generated, a system of field equations, mathematically represented by partial differential equations (PDEs) that describe the physics problem in concern, is applied over each element [17]. The fields within each element is characterized by simple linear or quadratic functions, governed by the element's type—the number of its nodes leading to a finite number of degrees of freedom (DOFs)—6 per node translations along and rotations about the x, y, z axes in structural analysis). Contributions of all elements are assembled into a large sparse matrix equation system which can be solved by a tailored numerical method, a simply a solver [17]. Figure 8.5a–c depicts a simple 2D structural analysis model with its problem definition including the geometry, its mesh structure and boundary conditions, as well as details of a single finite-element. In the following figure (Fig. 8.6), the FEM process workflow is given.

The *finite-volume method* (*FVM*) is similar in a way to FEM analysis' start by breaking down the object in concern into small finite-sized elements (called *cells*). These elements can be vertex or cell centered as illustrated in Fig. 8.7a, b. They can be structured as in FDM grids or unstructured. The method, common for dealing with fluid flow problems, is based on an idea that many laws governing physical phenomena are *conservation laws*, leading to an approach of a balance of inputs and outputs of each cell or flux conservation equations defined in an averaged way over the cells [17]. Figure 8.7b tracks the rate of increase of a quantify based on the difference between the flux of quantity in and the flux of quantity out without have a source inside the control volume. Table 8.2 compares the numerical methods covered in this section.

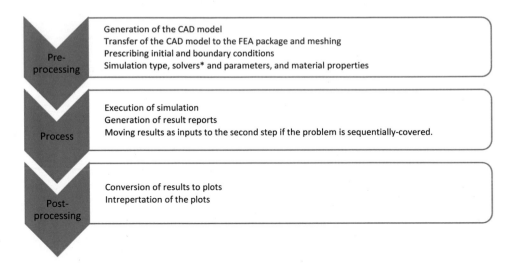

Fig. 8.6 FEM process flow (*: if multiple ones are available to choose from)

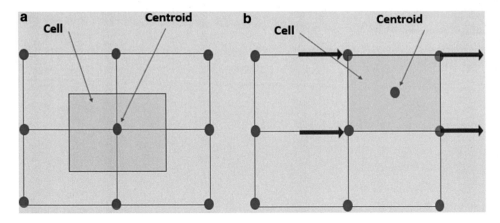

Fig. 8.7 **a** Structured vertex-based FVM control volume **b** Cell-based FVM control volume with their centroids (solution location)

8.2.2 CAE for Manufacturing Processes

CAE has long been used in analyses of engineered products, structures, and systems including machinery [18]. Its use in manufacturing engineering has been growing in the last couple of decades. Major manufacturing companies have been utilizing engineering numerical analysis software in design of parts, processes, and tooling, especially in an integrated fashion (*integrated manufacturing design* (IMD)) where all the three facets are handled concurrently. After accomplishing IMD, companies have been able to realize

Table 8.2 Comparison of numerical analysis methods [17, 21]

Numerical method	Advantages	Disadvantages
BEM	• Is often more efficient than other methods, including FEM, in terms of computational resources for problems where there is a small surface/volume ratio	• It works by constructing a mesh over the modeled surface. However, for many problems, it is significantly less efficient than the other three methods • Result in fully populated matrices meaning that the storage requirements and computational time will tend to grow according to the square of the problem size • Compression techniques (i.e. multipole expansions or adaptive cross approximation/hierarchical matrices) can be used to address these problems, though at the cost of added complexity and with a success-rate depending on the nature of the problem being solved and the geometry involved
FDM	• Easiest to implement if rectangular or prismatic shape geometry is broken into a regular grid • Regular grids can be used in very-large-scale simulations on supercomputers • Easy to increase element order	• Runs into problems handling curved boundaries for the purpose of defining the boundary conditions which are needed to truncate the computational domain • For cases that need high accuracy, the extra effort in making boundary-fitted meshes and the associated complications of such meshes for the implementation becomes critical • More difficult to use for handling material discontinuities • Does not lend itself for local grid refinement or anything similar to adaptive mesh refinement

(continued)

Table 8.2 (continued)

Numerical method	Advantages	Disadvantages
FEA/FEM	• A very general method for a variety of multi-physics analysis • Easy to increase the order of elements/resulting polynomials for getting more accurate results • Adaptive mesh refinement allows finer meshes where needed also improving the accuracy of the results • Mixed formulations (for a couple analysis like electromagnetic heating) are easy to handle • Curved and irregular geometries handled in a natural way	• Most difficult to implement due to complexity of the modeling required • In certain time-dependent simulations, explicit solvers are needed for reasons of efficiency. Implementing such solver techniques is more difficult for this method, when compared to the others
• FVM	• Is a natural choice in CFD problems • Implementation of it is comparatively straightforward • Curved and irregular geometries handled in a natural way • Only needs to do flux evaluation for the cell boundaries including nonlinear problems making it extra powerful for robust handling of nonlinear conservation laws in transport problems • The local accuracy of it can be increased by refining its mesh similar to FEM	• There are two viable alternatives to it—FEM and FDM • The functions that approximate the solution cannot be easily made of higher order, making it FVM's main disadvantage, when compared to the other two

consequent part and tool manufacturing via different methods including CNC machining and 3D printing. CAE tools for manufacturing have become available to smaller companies with the advents in computers and reduction in their costs, even though the cost of some of these software tools are still not within the reach of most small and medium size manufacturers. Increase in the utilization of CAE in IMD resulted in the need for introducing these tools to engineering graduates. Schools have been raised up to the challenge, and various courses have been designed and offered. Cost of software is not an issue for academic institutions since many software companies have educational programs offering drastic price cuts to encourage schools to integrate their software with their manufacturing curriculum [18].

There are multiple ways that these CAE tools can be involved in part, associated manufacturing process and tooling analysis, utilizing multiphysics FEM tools, pre-simulation tools, machine simulators, geometric analysis tools, along with others:

Multiphysics FEM tools:

- *Analyzing the strength of a product or system design* after the completion of the initial design or during continuous improvements efforts: FEA tools like SIMULIA or Ansys can be employed in studying strength of a component like a connecting rod in an internal combustion engine during its operation or, durability of a complete car structural model is tested in PAM/Crash software [22].
- *Analyzing the feasibility of a manufacturing process* such as forging by employing FEA tools like DEFORM software to mimic the process along with its part (i.e. a connecting rod), and associated die design to achieve IMD: [23]
  - FEA-based simulation tools like ProCAST [24] has the ability to analyze some of the main casting applications for optimizing them, including sand-casting, die-casting, lost foam and investment casting, continuous casting, and centrifugal casting processes. This software can incorporate die-casting machines into the process simulation, use advanced porosity models when needed, and study grain structures of resulting castings. It can also analyze microstructures after heat treatment.
  - In a similar fashion Moldflow and Solidworks Plastics Analysis [25] can handle injection molding process filling system design (sprue, runner, and gate), along with shrinkage analysis, analysis of meld and weld lines, warpage, and cooling circuit design and analysis—Figs. 8.8a–d and 8.9a, b [25].

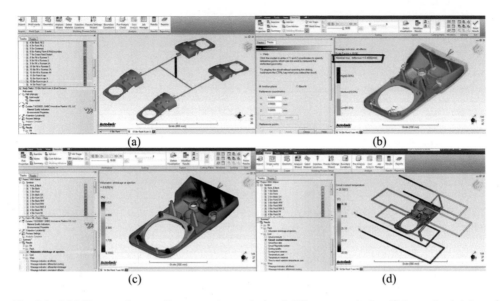

Fig. 8.8 Moldflow analyses **a** Runner and gate design **b** Warpage analysis **c** Volumetric shrinkage analysis **d** Cooling circuit analysis [25]

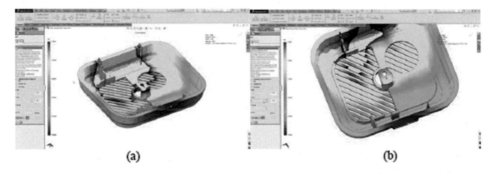

Fig. 8.9 Solidworks plastics—standard software **a** cavity fill analysis **b** weld line analysis [25]

Pre-simulation tools:

- Ohio State University's CastView is based on *voxel analysis (onion peeling method)* for thin/thick sections of castings, and also handles *ray patterns mimicking high-pressure* cavity filling for filling analysis. It is considered as a pre-simulation tool allowing the designer to make important changes in quick turnaround before a complete FEM can be run analyzing the die-casting process. The designer can determine the number of gates and their locations, vents and their locations, and take precautions for filling thin and thick sections of the part [18].

Machine simulators:

- These tools are mainly intended for training students in machine use like an injection molding machine or a CNC—Fig. 8.10a–c [25]. However, some of these tools may have an optimization feature that gives results of process parameters selected by the users in the form a performance review, mimicking a complete process analysis.

Geometric analysis:

- CAM tools like Mastercam can through their *verification feature* analyze and check if a cutting tool operation will yield its expected results without gouging the stock,

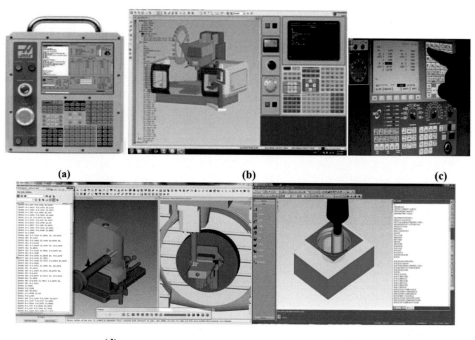

(a) (b) (c)

(d) (e)

Fig. 8.10 **a** HASS dual system control simulator **b** Swansoft CNC simulation software with a HAAS CNC control panel being shown **c** Swansoft software running on a tablet **d** VERICUT CNC software in verification mode **e** VERICUT in machine probing mode [25]

similar to what is shown in Fig. 8.10d. Geometric analysis is done by subtracting the volume of the cutting tool intersecting with the sections of the stock. Vericut CNC tool can, in addition to verification, do simulation of the *probing process*—Fig. 8.10e [25].

- 3D Printing software like Cura can do *part/process analysis* estimating printing process times and costs as well as optimization of support structures, layers and part quality.

Other tools may include material and manufacturing process selection databases and software, dimensional and tolerance analysis, product life-cycle management and sustainability software that includes eco-audits, and cost analysis and estimation tools, just to name a few.

The lead author also looked into understanding capabilities of the some of the tools mentioned in the multiphysics-based FEM and pre-simulation tools with a simple criterion that included [26]:

- Application areas—coverage of different i.e. casting processes.
- Ease of use.
- Time requirement for completing the analysis (including pre-, simulation, and post-steps).
- Geometry modeling capability—geometry formats acceptable to the software tool and its design ability.
- Presentation of results—do they require interpretation?
- Accuracy of the results.

In Practice

Following exercises were originally used in a graduate level, Manufacturing Processes and Simulation, course at Ohio State University [18] focusing on a specific forging process called *upsetting*. The process involves placing a solid cylindrical *billet* between two flat dies and reducing the workpiece's height under compression. Cold version of the upsetting process typically employed for testing material's behavior at large strains, complementing the traditional tensile test that is limited to small stresses due to necking. At low temperatures, many metals exhibit strong work-hardening behavior, leading to flow-stress relationship in power form. The *ring test* (based on upsetting a ring-shaped specimen) is also employed to measure the friction factor at the die and specimen interfaces, determining the material flow and final shape of the specimen. If the friction at the interfaces is zero, both the inner and outer diameters of the ring expand as if it were a solid disk. With increasing friction, the inner diameter becomes smaller. For a particular reduction in height, there is a critical friction value (m) where the internal diameter increases from its original dimension if m is low, and decreases if m is high. By measuring the change in the specimen's internal diameter, and using curves obtained through theoretical and numerical (FEM) analyses, we can determine the coefficient of friction or friction factor. Each ring geometry has its own specific set of curves. The most common specimen geometry has outer diameter to inner diameter to height proportions of the specimens 6:3:2 while the actual size of the specimen usually is not relevant. Thus, once you know the percentage reductions in internal diameter and height, you can determine the coefficient of friction or friction factor from the appropriate chart.

The objective of this exercise is to introduce the different upsetting processes and the process conditions to the students. The exercise is divided into two cases to simulate cold and hot upsetting processes and cold ring tests. You can use the DEFORM or a similar software to complete the following:

Step 1: Upsetting (isothermal)

Step 1 involves cold and hot upsetting of cylindrical billets between two flat rigid dies at different friction conditions. You can choose the material to simulate along with its process temperature.

Geometry:

Initial Radius for Workpiece: 1.0 in
Initial Height for Workpiece: 4.0 in
Mesh for Workpiece: 250 elements
Die Diameter: 10.0 in

Process Conditions:
Upper Die Speed: 1.0 in/s
Lower Die Speed: 0.0 in/s
Temperature: Please choose your process temperature
Reduction in Height: 50%
Friction factor: m = 0.0 and 1.0 (no friction and sticking friction)
Number of steps: 100

Assignment

(a) Prepare the following the plots: (for both cold and hot working)
   - For both friction factors, a deformed mesh plot at maximum reduction.
   - For both friction factors, an effective strain contour plot at maximum reduction.
   - A load-stroke curve for both friction factors.
(b) Answer the following questions briefly.
   - What influence does the friction factor have on the deformed geometry?

- Where does the maximum effective strain occur? Compare the effective strain contour plots for both friction factors.
- How is the load-stroke curve affected by the friction factor. Why does the curve not start at the origin? (Hint: Think about how DEFORM or its substitute models the material behavior).
- If the diameter of the workpiece was an inch, what major problem can be expected during upsetting? Discuss.
- Compare the results within the group for different working conditions, and discuss the influence of working temperature on the load-stroke curve.

Step 2. Simulation of a Ring Test (Cold only)

The objective of this step is to simulate the compression of a ring using flat, rigid dies under isothermal conditions. The workpiece material is the same as Step 1.

Geometry:

Initial Outer Diameter for Workpiece: 6.0 in
Initial Inner Diameter for Workpiece: 3.0 in
Initial Height for Workpiece: 2.0 in
Mesh for Workpiece: 250 elements
Die Diameter: 10 in

Process Conditions:

Upper Die Speed: 1.0 in/s
Lower Die Speed: 0.0 in/s
Stroke: 1.0 in
Temperature: 70 F
Friction Factor: m = 0.0, 0.1, 0.3 and 1.0

Number of simulation steps: 100

Assignment:

(a) Prepare the following plots with your report:
   - For all four cases a deformed mesh plot at maximum reduction.
   - For the first and fourth case, a plot of effective strain contours.
   - For all four cases, a load-stroke curve obtained by FEM.
   - Using the data from results of DEFORM (or its substitute) analyses of 10, 20, 30, 40 & 50% reductions in height, obtain a curve of % reduction in internal diameter versus the % reduction in height for all cases.
(b) Answer the following questions briefly.

– What effect does friction factor m have on the final shape of the ring and on the value of the final internal diameter?
– What effect does friction factor m have on the effective strain at the inner diameter and mid height area?
– For the first case, theoretically predict the effective strain at maximum reduction.
– For the fourth case, check if you have zones of low effective strain. Can you explain the formation of these zones?

8.2.3   CAE for Manufacturing Systems

As mentioned earlier in Chap. 1, CAE is also employed in understanding many aspects of manufacturing systems. This type of an effort may encompass tracking and analyzing material, part/product, worker flows, storage and machine utilization via discrete-event simulation tools (including in 3D), safety and health hazard analysis with ergonomics or human modules. Additional tasks include but are not limited to generation and execution of CNC and robot programs via CAM and off-line programming abilities, or PLC programs controlling various parts of a digital factory or a whole plant. Figure 8.11 illustrates a case of *body in white (BIW)* robot assembly example where robot programmers can simulate and validate their programs in 3D Experience/Delmia [27].

The lead author has used off-line programming tools such as ABB's RobotStudio and FANUC's SimPro/Roboguide including its HandlingPro and PalletPro modules. He also

Fig. 8.11 The 3D Experience/DELMIA model of a manufacturing facility [27]

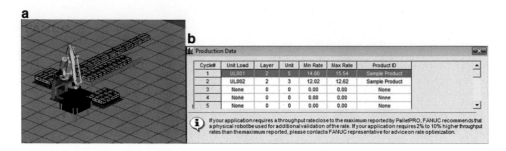

Fig. 8.12 **a** FANUC PalletPro palletizing exercise **b** Production data

has had experiences with DDS' Delmia modules for robotics and human (ergonomics), and worked with his students to define new *robot configurations* (*dynamic models of those robots*) in DDS' CATIA to be employed in Delmia. At the time, CATIA ran on the background of DELMIA. After DDS migrated their tools to a cloud-based environment, 3D Experience, he prepared spot welding programs with his students for ABB robots. Figure 8.12a is presenting a FANUC PalletPro programming/simulation exercise, employed as a laboratory assignment for the lead author's ENGR 4700 Robotics and Automation course. After building the robotic palletizing cell, students run a simple simulation that determines the progress of the palletizing process for two different pallet set-ups. The students can see the process details (robot speeds and box placement patterns in a pallet, as well as the roles of separator sheets) and record the process determining how long it takes to complete it—Fig. 8.12b.

According to DDS, 3D Experience aims for "as a unified collaborative environment bringing together all participants and enables secure, real-time collaboration on any device. Its user experience brings web-style simplicity and speed to individuals, teams and departments even when dispersed across different locations." [28] Besides SOLIDWORKS, CATIA, and SIMULIA, DDS offered the following capabilities for systematic/enterprise-wide solutions at the time this book was being written:

- 3DEXCITE provides comprehensive solutions for collaboration and 3D product content creation. Thus, the software helps global brands to create marketing experiences where the customers can interact with the products through rich media [29].
- 3DVIA helps consumers make the best buying decisions, i.e. home improvement retailers and brand manufacturers can use this cloud-based omnichannel space planning tool to generate high quality sales leads in short cycles [30].
- BIOVIA is committed to enhancing and speeding innovation, increasing productivity, improving quality and compliance, reducing costs and accelerating product development for customers in biological, chemical, and materials industries. The software involves the diversity of science, experimental processes and information requirements,

end-to-end, across research, development, *Quality Assurance (QA)/ Quality Control (QC)* and manufacturing [31].

- Centric PLM allows companies to get closer to their customers by becoming rapidly responsive to the market demand changes, accelerating time to the market by handling marketing, financial, and legal requirements, increase productivity by eliminating waste of time searching for data and files, ensure quality and sustainability by making sure that products meet global standards before they are released to the markets [32].
- DELMIA enables industries and service companies to model, control, and optimize their operations [33].
- ENOVIA provides a secure and real-time environment for finding, sharing, reviewing and reporting information, help manage the product planning processes and programs including customer requirements for successful strategic customer relationships along with suppliers, assuring quality and reliability via *Quality Function Deployment (QFD)* and *Failure Mode Effects Analysis (FMEA)* [34].
- According to DDS, leading companies worldwide rely on NETVIBES data discovery tools to search, reveal, and manage their information needs/assets for quick and effective decision making, along with unified data access and improved productivity [35]. Capabilities of this tool include digital twin experiences or AI-powered text analytics software, Proxem Studio.

We can deduce that PLM and product data management, project management, e-commerce and strategic customer relationships tools have taken the center stage within the DDS tool kits, as they are very critical to any manufacturing enterprise's business. One of the interesting tools available is GEOVIA which models and simulates earth, our environment, and its major destructors including mining and urbanization. The software has the tools for sustainable capture, and use, and re-use of our planet's resources [36].

8.2.4 Digital Twins

Chapter 1 includes an introduction to digital (also called *virtual*) twins and digital thread areas. This chapter will mainly focus on digital twins and covering different examples than what was covered earlier. Software companies like Ansys [37] or DDS [38], and industrial companies with major software and control presence like Siemens [39] have been focusing on digital twins.

According to a joint Accenture and DDS study [38], digital twin is "a real time virtual representation of a product, process, or a whole system that is used to model, visualize, and provide feedback on properties and performance, and is based on an underlying digital thread". The digital thread is the network of operations and associated digital tools creating, communicating, and transacting product information along the product's lifecycle, allowing the digital twin to be continuously updated based on the performance of its physical twin (asset in concern) [38]. Better understanding of the physical twin via IoT improves its modeling, or its digital counterpart, and in turn control of itself during its use. This is true for the product's life cycle including its development, manufacturing, and dismantling at the end of its life as shown in Fig. 8.13. Information generated by computation can be applied to improve the development and manufacturing steps via innovation, and also help optimize the use of the product and its dismantling at the end of its life.

Table 8.3 below lists the capabilities of digital twins employed in manufacturing enterprises. Impact of these capabilities are be summarized below [38]:

- Product cost reduction
- Operation and product footprint reduction
- Reduction in response to the market
- Regulatory and EHS risk reduction
- Enablement of cross-functional collaboration
- Enablement of new service models.

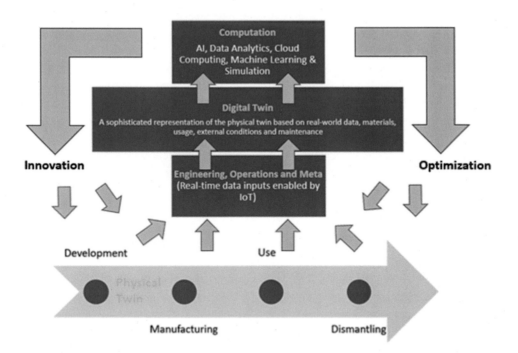

Fig. 8.13 Digital twin process flow [38]

Table 8.3 Key capabilities of digital twins in a manufacturing enterprise [38]

Research and development	Design and engineering	Manufacturing	Transport and logistics	Product use	Decommission and end-of-life
– Access to past product performance data – In-silico development of new materials – Multidisciplinary research collaboration – Organized access to relevant data – Visibility of lifecycle impact	– Access to past product performance data – Generative design – In-silico prototyping and testing – Multidiscipline design collaboration – Organized access to relevant data – Visibility of lifecycle impact	– New product trials and ramp-up – Improved operational feedback at the point of worker interaction – Intelligent monitoring and maintenance of equipment – Manufacturing process simulation and optimization – Plant facility layout simulation and improvement – Plant and machinery controls automation based on real-time operations	– Algorithmic planning and route optimization – Container tracking and management – Fleet management – Sensor-based shipments condition monitoring – Virtualization and visualization of logistics facilities and infrastructure – Virtualization and visualization of logistic networks	– Advanced failure warning and risk management – Over the air software performance optimization – Intelligent asset service and maintenance – Real-time generation of operational improvement insights – Remote asset monitoring and diagnostics	– Data-based decom planning/life-extension assessment – Decommission process simulation and planning – Decommission execution simulation and planning – Detailed visibility into asset and component status, material composition and design – Material and component recovery tracking

8.2.4.1 Digital Twin Use Cases

Two use-cases are presented in this section, a Tesla Model 3 breaking issue, and Sanofi's Framingham lighthouse facility [38].

The first case in concern is about digital twin technology helping improve a product, Tesla Model 3, with after-market (sales) asset performance optimization [38]. Consumer Reports (CR) magazine in May 2018 reported braking, controls, and ride quality issues for the car, and did not recommend it to its readers [40]. The following weekend, Tesla sent out an over-the-air software update, improving the car's 60 mph stopping distance by 19 feet (cutting it to 133), a value which is about average for luxury sedans (varying within a range of 120–140 feet). Soon after, CR updated its review and gave the car its approval. The over-the-air update and improvement was the first for Jake Fisher, the director of auto testing at CR, who had tested more than 1000 cars in his career at the magazine. Tesla's operational digital twin capability is a true representation of reality, since it can acquire the mileage from its cars across different locations under different wind conditions and calibrate their digital twins running aerodynamics and drag coefficient analysis [38].

Sanofi's Framingham production facility is the focus of the second case. It is a fully digitalized, continuous production facility where R&D has direct impact on the production processes [38]. Digital twin technology, along with real-time data capture and analysis, is employed to optimize the processes. The whole production effort is paperless and is 80 times more productive than that of a similar traditional facility. This system can also make medicine in less lead times for twice the number of patients and all happens in a smaller environmental footprint: Following performance indicators were observed for Sanofi's new system:

- 80% reduction in energy consumption and CO_2e emissions per year,
- 91% reduction in water footprint,
- 94% reduction in use of chemicals,
- and 321 tons of waste reduction annually.

Review Questions

1. What is the reason for forcing today's engineers to be versatile?
2. Define the term "Internet of Things (IoT)" and list some of its applications in automotive manufacturing.
3. What is the difference between IoT and "Industrial Internet of Things (IIoT)"?
4. What are the critical factors for understanding IoT technology? Elaborate each one briefly.
5. List the layers of the IoT Architecture and explain the process flow.
6. Discuss the impact of IoT on today's manufacturing.
7. Give present and future of three IoT technologies, edge computing, voice AI, and vision AI.

8. Make a mind-map encompassing the major analysis types in multi-physics simulation.
9. What is the boundary element method? Give an application example where it is effective.
10. What is the finite difference method? Give an application example where it is effective.
11. What is the finite element method? Give an application example where it is effective.
12. What is the finite volume method? Give an application example where it is effective.
13. Explain in detail the FEM process flow with an example multiphysics problem.
14. What is IMD? What is its main benefit?
15. What are two ways that the FEM-based CAE tools are utilized in manufacturing process simulation?
16. Elaborate on the role of pre-simulation tools in CAE for manufacturing processes.
17. Elaborate on the role of machine simulators in process improvements.
18. What does geometric analysis do in CAM environments?
19. What are the major functions of a 3D printing software like Cura?
20. What factors can be used in selecting an FEM or pre-simulation tool?
21. What are the major analysis types when employing a CAE analysis for manufacturing systems tool? (Hint: Use DELMIA or SIEMENS PLM software as a reference)
22. What are the key capabilities of digital twin applications?
23. What are the impacts of the key capabilities mentioned in the previous question on a manufacturing enterprise?

Research Questions

1. If you've ever used an app to turn your lights on at home, you've used the application layer of the IoT application to do so. Please explore and analyze such an application by explaining each layer.
2. Research Lexmark's new Optra IoT Platform to see its capabilities.
3. Find a simple BEM analysis case and summarize its critical components—geometry/mesh, initial and boundary conditions etc.
4. Study implicit, explicit, and Crank–Nicholson schemes used in FDM analysis. Complete one example for each.
5. Research capabilities and components of the software DEFORM used in modeling forming process. Elaborate on each briefly.
6. Review an FEM study case of IMD. Present the FEM model's components (including its geometry, mesh selection, initial condition, boundary conditions, process and simulation parameters).
7. Research capabilities of material selection software like Ansys Granta EduPack.
8. Research modules and their capabilities of FANUC's SimPro software. Explain each one briefly.

9. Conduct a literature review to find a case study on digital twin applications in manufacturing, and summarize it in a single PowerPoint slide.
10. Conduct a literature review to find a case study on digital thread applications in manufacturing, and summarize it in a single PowerPoint slide.
11. FANUC America recently unveiled its ROBODRILL's visual twin at a demo held at JTEKT Toyoda America Corp.'s Open House in Arlington, Heights, IL. What is a visual twin and how does it relate to its digital twin?

Discussion Questions

1. Discuss the present state of IoT, and its impact on today's manufacturing.
2. Discuss the future of IoT.
3. Discuss use of FEM vs FDM in a 1D thermal analysis (in a wall) where heat conduction occurs within the object in concern, with the boundary conditions of convective heat transfer on both sides of the object. Which method do you choose and why?
4. Generate a sketch of an FEM model for analyzing closed-die forging. Include the geometry (of your choice including its simplification if needed), possible mesh structure, initial, process and boundary conditions—based on your assumptions)
5. Make three comprehensive mind maps summarizing all computing tools used in manufacturing engineering including (processes, systems, and other analysis).

References

1. Sirinterlikci, A., Al-Jaroodi, J., Kesserwan, N. (2020), Unpublished RMU Industrial Internet of Things (IIoT) course proposal.
2. Neptun Glass srl, located at: https://neptunglass.com, accessed September 20, 2022.
3. IoT Architecture, located at: https://www.celona.io/network-architecture/iot-architecture, accessed September 20, 2022.
4. Stage IoT Architecture, located at : https://www.google.com/url?sa=i&url=https%3A%2F%2Fwww.geeksforgeeks.org%2Farchitecture-of-internet-of-things-iot%2F&psig=AOvVaw0RNwYpNlqlnU-7kbNoVTRO&ust=1665157910045000&source=images&cd=vfe&ved=0CAsQjRxqFwoTCPCIxt_6y_oCFQAAAAAdAAAAABAE, accessed September 20, 2022.
5. Explore IOT Kit, located at: https://Arduino.cc/Explore-IOT-Kit, accessed September 20, 2022.
6. Oplà, located at: https://Opla.Arduino.cc/, accessed September 20, 2022.
7. Gosh, I., (2021). 4 key areas where AI and IoT are being combined, located at: https://www.weforum.org/agenda/2021/03/ai-is-fusing-with-the-internet-of-things-to-create-new-technology-innovations, accessed September 20, 2022.
8. Industry 4.0 Mind Map, located at: https://intelligence.weforum.org/topics/a1Gb0000001RIhBEAW?tab=publications, accessed September 20, 2022.

9. Globe newswire news-release -01/13/2022, located at: https://www.globenewswire.com/news-release/2022/01/13/2366783/0/en/Global-Internet-of-Things-IoT-Market-Size-To-Hit-USD-1-842-Billion-by-2028-at-a-24-5-CAGR-Growth-with-COVID-19-Analysis-Facts-Factors.html, accessed September 20, 2022.

10. Future IoT, located at: https://www.ericsson.com/en/future-technologies/future-iot, accessed September 20, 2022.

11. Emerging Internet of Things (IoT) Trends in 2022 - Mastek Blog, located at: https://blog.mastek.com/emerging-internet-of-things-iot-trends-in-2022, accessed September 20, 2022.

12. Aiot-when-ai-meets-iot-technology, located at: https://visualcapitalist.com/aiot-when-ai-meets-iot-technology, accessed September 20, 2022.

13. FEATool Multiphysics, https://www.featool.com/cfd-toolbox/, accessed at September 20, 2022.

14. Finite element analysis of biological soft tissue surrounded by a deformable, located at: https://tbiomed.biomedcentral.com/articles/https://doi.org/10.1186/s12976-018-0094-9/, accessed September 20, 2022.

15. Sivaiah, S., Murali, G., Reddy, C.G.K. (2012), Finite Element Analysis of Chemical Reaction and Radiation Effects on Isothermal Vertical Oscillating Plate with Variable Mass Diffusion, located at: https://www.hindawi.com/journals/isrn/2012/401515, accessed September 20, 2022.

16. Boundary Element Model Course Notes, https://web.stanford.edu/class/energy281/BoundaryElementMethod.pdf, accessed September 20, 2022.

17. Sjodin, B., (2016). What's difference between fem, fdm, and fvm https://www.machinedesign.com/datasheet/what-s-difference-between-fem-fdm-and-fvm-pdf-download, accessed September 20, 2022.

18. Sirinterlikci, A., Habib, S.G., (2003), Utilizing Manufacturing Process Simulation Tools as Instructional, American Society for Engineering Education (ASEE) Conference and Exposition, Nashville, TN.

19. Fundamentals of fea meshing for structural analysis/, located at: https://www.ansys.com/blog/fundamentals-of-fea-meshing-for-structural-analysis/, accessed September 20, 2022.

20. Tekkaya A.E., Soyarslan, C. (2019), *Finite Element Method - CIRP Encyclopedia of Production Engineering* pp. 508–514, located at https://link.springer.com/referenceworkentry/https://doi.org/10.1007/978-3-642-20617-7_16699, accessed September 20, 2022.

21. Finite Element Method (FEM) vs. Finite Volume Method (FVM) in Field Solvers for Electronics, located at https://resources.pcb.cadence.com/blog/2020-finite-element-method-fem-vs-finite-volume-method-fvm-in-field-solvers-for-electronics, September 20, 2022.

22. Pam-Crash – ESI Group, located at: https://www.esi-group.com/pash-crash, accessed September 20, 2022.

23. Deform - Scientific Forming Technologies Corporation, located at: https://www.deform.com/, accessed September 20, 2022.

24. Casting Simulation Software – ESI Group, located at: https://www.esi-group.com/products/casting, accessed September 20, 2022.

25. Ertekin, Y., Chiou, R. (2015), Integration of Simulation Tools in Manufacturing Processes Course, American Society for Engineering Education (ASEE) Conference and Exhibition, Seattle, WA.

26. Habib, S.G. (2001), Comparison of Geometric Visualization and Finite Difference Method in Analyzing Casting Processes, Master's Thesis, Texas Tech University.

27. Delmia Robotics, located at: https://4dsysco.com/delmia/delmia-robotics, accessed September 20, 2022.

28. General reference on 3D Experience.

29. Engineer the excitement I 3DEXCITE, located at: https://www.3ds.com/3dexcite/, accessed September 20, 2022.

30. Shape Your Dream | 3DVIA, located at: https://www.3ds.com/products-services/3dvia/, accessed September 20, 2022.
31. Model the Biosphere | BIOVIA, https://www.3ds.com/products-services/biovia/, accessed September 20, 2022.
32. Centric Software, located at: https://www.centricsoftware.com/, accessed September 20, 2022.
33. Global Operations Software – DELMIA – Dassault Systemes, located at: https://www.3ds.com/products-services/netvibes/, accessed September 20, 2022.
34. ENOVIA | TECHNIA, located at: https://www.technia.us/, accessed September 20, 2022.
35. NETVIBES Products by Dassault Systemes, located at https://www.3ds.com/products-services/nevibes/products/, accessed September 20, 2022.
36. Model the sustainable planet | GEOVIA – Dassault Systemes, located at: located at https://www.3ds.com/products-services/geovia/, accessed September 20, 2022.
37. MacDonald, C., Dion, B., Davoudabadi, M., Creating a Digital Twin for a Pump, ANSYS Corporate Resource Library, located at: https://www.ansys.com/-/media/ansys/corporate/resourcel ibrary/article/creating-a-digital-twin-for-a-pump-aa-v11-i1.pdf, accessed November 4, 2019.
38. DESIGNING DISRUPTION: The critical role of virtual twins in accelerating sustainability, located at: https://www.3ds.com/sustainability/designing-disruption, accessed September 20, 2022.
39. Maturing simulation and test capabilities digital twin, located at https://resources.sw.siemens.com/en-US/white-paper-maturing-simulation-and-test-capabilities-digital-twin?stc=usdi10 0005/, accessed September 20, 2022.
40. Tesla Model 3 Falls Short of a CR Recommendation - Despite record range and agile handling, issues with braking, controls, and ride quality hurt the Model 3's Overall Score, located at: https://www.consumerreports.org/hybrids-evs/tesla-model-3-review-falls-short-of-consumer-reports-recommendation/, accessed September 20, 2022.

Reverse Engineering

9

According to Dym and Little [1], "*reverse engineering is examining competitive or similar or prior products in great detail by dissecting them or literally taking them apart.*" It is the scientific method employed by developers to understand a product's structure and its inner workings for developing competing ones [2]. Reverse engineering (RE) is also an invaluable tool used by researchers, academics, and students in many disciplines, who reverse engineer technology to discover, and learn from, a product or system's structure and design. Through reverse engineering, developers try to gain insight into their own design problem by looking at how it was addressed by others. However, there will be some restrictions [3] to reverse engineering efforts including:

- Designs which are in concern may be expensive to duplicate, and thus not be feasible.
- Designs may be protected by the intellectual property laws, namely patent or copyright laws.
- Designs may belong to a close competitor, thus a licensing agreement for use of their intellectual property may not be attainable.
- Designs may not be as useful.

Reverse engineering has two major thrusts: its *methodology* including the means used in studying the design in concern, and the *technology* utilized in accomplishing the first goal. Both are the subjects of the following sections of this chapter.

© The Author(s), under exclusive license to Springer Nature Switzerland AG 2023   251
A. Sirinterlikci and Y. Ertekin, *A Comprehensive Approach to Digital Manufacturing*,
Synthesis Lectures on Mechanical Engineering,
https://doi.org/10.1007/978-3-031-25354-6_9

9.1   Reverse Engineering Methodology

Commonly, RE is performed using the *clean-room or Chinese wall* method [4]. RE is conducted in a sequential fashion in that method. A team of engineers are tasked to disassemble the product to study and describe what it does in as much detail as possible at a high level of abstraction. Description prepared by the first team is given to another team who has no previous knowledge of the product. The second team then builds a product from the given description. This product might achieve the same result but will probably have a different solution approach, helping circumvent the intellectual property (IP) protections [5].

Reverse engineering methodology helps developers to answers what, how, why questions about the design in concern alike [3]:

- What does it do?
- How does it do that?
- Why would you want to do that?

Figure 9.1 depicts the reverse engineering process which is composed of three stages, *reverse engineering that includes investigation, prediction, and hypothesis building/concrete experience, modeling and analysis of the design in concern, and redesign or reengineering for one's own design.*

Reverse engineering stage starts with the investigation, *prediction, and hypothesis building* step as shown in Fig. 9.1 [3]. This first step involves developing *a black box model* of the product by analyzing it at a high level. The black box model is the graphic representation of the design in concern with its inputs entering the box from the left and its outputs leaving on the right. For shake flashlights as illustrated in Fig. 9.2, shaking

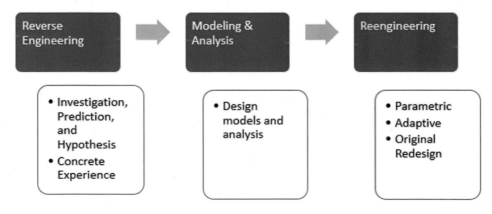

Fig. 9.1 Reverse engineering [3]

Fig. 9.2   Black box model of a shake flashlight

motion and the power switch serve as the inputs. When in operation, flash light produces light without heating since it includes LEDs. Thus, its sole output becomes the light if we also ignore the noise it makes while shaking. The seconds step of the reverse engineering stage is using or experiencing the product as if a consumer would. The lead author asks his students to see if the performance of their chosen product matches its claims. This also includes checking if the product actually works. In the case of Dinoco Cruz Ramirez "Revvin' Action" Windup Toy shown in Fig. 9.3a, the student team found that its performance was lacking in terms of travel distance with 40 feet of travel with 5 activation motions, when compared to the claim of 50 feet. They observed that the toy increased its speed after initial movement, and followed a straight path. They also found the user age limit of 4+ appropriate due to small parts not being accessible from outside, and its resale price of $10.99 reasonable when benchmarked against similar toys. In the case of a Black and Decker hedge trimmer (Fig. 9.3b), a more detailed analysis was conducted by Gary Kinzel and his team after looking into the product performance attributes like shearing speed of 3300 strokes/min, motor rotation of 16,500 rpm, and gear reduction of 5:1 [3]. Following Table 9.1 summarizes their analysis and presents a customer review example they included in their presentation.

Investigation, prediction, and hypothesis building step can be completed by *listing assumed working principles of the product, stating process description or an activity diagram*, and *studying economic feasibility of redesign*. At this instant of the process *a patent search* can be carried out for understanding the inner workings of the product in concern

a   **b**

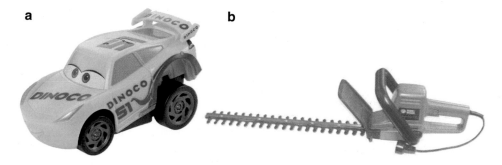

Fig. 9.3   **a** Dinoco Cruz Ramirez, **b** a Black and Decker hedge trimmer [3]

Table 9.1 Further understanding a hedge-trimmer [3]

Questions?	Answers
What is the market of this product?	• "Suitable for small shrubbery"—Black and Decker Product Catalog • Homeowners with small yards and limited budget • For use only 3–4 times a year
What are the costs associated with this product?	• Design—Manufacturing—Assembly—Packaging—Resale—$40.00
How long will this product last?	• Assumed durability of each component (outdoor use, dirt) • Availability of replacement parts and service shops
What features does this product have that are important?	• Molded-in cord retainer • Lock off switch prevents accidental start-up • Lock on switch for continuous running • Lightweight design for less fatigue of the user—4.5 lbs
What are the customers who bought this product think?	By: jenni796 (Fri Apr 7'00) Pros: light weight, very durable Cons: none • Trimming the bushes is my only contribution to our 2 acre yard • I bought my first Black & Decker hedge trimmer at Wal*Mart because it was very inexpensive compared to most other trimmers • Black and Decker has an excellent reputation • The 13″ seemed a little too small… The 18″ seemed heavier • I also wanted electric rather than gas because being a busy woman, I had no time to learn about mixing gas Durability: Excellent Noise Level: Average Purchase Price: $25.00

by visiting the United States Patent Trademark Office (USPTO) website [6] or the Google Patent Search [7]. Such a search may yield at least two patents pertaining to those hedge trimmers—Patent # 5,581,891 (1996) Hedge Trimmer with Combination Shearing and Sawing Blade Assembly, and Patent # 5,778,649 (1998) Power Driven Hedge Trimmer.

After the completion of investigation, prediction, and hypothesis building, reverse engineering stage continues with *concrete experience* efforts. The team will plan and execute the *disassembly of the product,* breaking into its parts revealing its hierarchy (System Level Design), in a way that their interactions are not lost during the process. The disassembly process can be documented with photographs (including *exploded disassembly photographs*) as shown in Fig. 9.4, sketches, and videos, especially if the product will be reassembled after the effort. During the disassembly processes, teams are aware that each system like a toy car is made from its *components* and their *connections (interactions),* becoming subassemblies and the overall product together. Thus, they study each component with its form (including dimensions) and functions. The connections or interactions amongst the components need to be determined and can include both the *intended* or

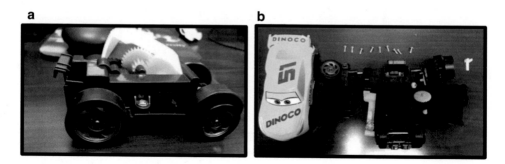

Fig. 9.4 Dinoco Cruz Ramirez disassembly process. **a** a subassembly, **b** full disassembly with all of its components

Fig. 9.5 Glass box model of the hedge trimmer with its mechanical and electrical interactions are marked with solid and dashed lines respectively [3]

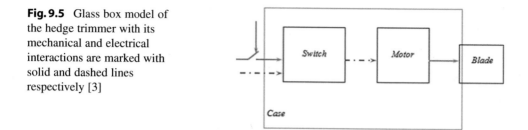

incidental ones like vibration and heat transfer due to the proximity of the components. *Flow charts, free body diagrams, and engineering specifications and their metrics can also* be produced. A *Glass Box Model* of the hedge trimmer is shown in Fig. 9.5. A diagram like a glass box represents the hierarchy of the products' subassemblies and components (electrical, mechanical, or task-based classification), their interactions amongst themselves (intentional/unintentional, flow of forces, wires, signals/data). Materials, manufacturing process information, as well as engineering specifications are also added to the glass box model after more questions are generated and answered during the process (Figs. 9.6 and 9.7). The RE team can also experiment with the system components or subassemblies. In the case of the hedger trimmer, Engineering specifications include the transfer of torque from the electrical motor via the gear box (gear head through its gears, input/output gear), associated rotational speeds and acceleration, in addition to the resulting forces [3]. In terms of the materials and manufacturing processes, following information was determined by the Kinzel's team even though it is at a higher level than expected since it does not explicitly identify what type of plastics are used [3]:

- Switch—Plastic/Injection Molded
- Gear—Steel/Die-Cast Steel
- Case—Plastic/Injection Molded

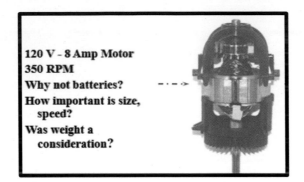

Fig. 9.6   Electrical motor and switch mechanism for the hedge trimmer [3]

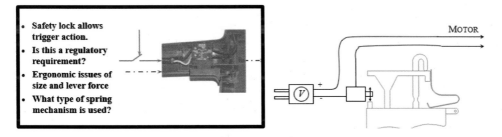

Fig. 9.7   A switch mechanism b. its sketch/circuit for the hedge trimmer [3]

- Handle—Plastic/Injection Molded
- Guard—Plastic/Injection Molded.

Modeling and analysis stage is the second stage of the process and made of two steps: *design modeling* and *design analy*sis [3]. In the first step, the team starts the *design or the redesign effort* based on what it learned from the previous RE stage and have a better idea on what can improve the existing design. Considering the physical principles involved, the patent restrictions, and the needs/wishes of the customers, the team then creates engineering models and associated metric ranges, and follow through by building alternative prototypes and test them with the parameters. The second step is about analyzing or simulating for optimization of the design after the team calibrates its engineering model based on the experiments carried out [3].

The third stage of the process is *reengineering or completion of the redesign process which* started in the previous *stage* and can have three different pathways (options) [3]:

- A *parametric redesign* is an evolution of the current design. New design variables (i.e. for a more robust mechanism or a stronger motor) are determined along with their acceptable ranges. Involves sensitivity analysis and tolerance design as well as prototyping and testing it.
- *Adaptive redesign* does not modify the current design but adds onto to satisfy a new function (i.e. a new subroutine to an existing computer routine). It involves searching for innovation solutions including new component configurations and further analysis, complemented by building and testing prototypes.
- An *original redesign* starts from scratch using the concepts, physical principles and outcome of the RE process to create a totally new product idea, replacing the current one. New design alternatives are often inspired by the inner workings of products, their materials, and how they were made. Again, a list of acceptable variable ranges is determined after choosing an alternative, and the new prototype is tested for improvements.

9.2   Reverse Engineering Technology

Reverse engineering technology encompasses all tools utilized in the reverse engineering process, including both the *hardware* and *software*. The hardware tools can be either *complex* or *simple, automatic* or *manual in nature, and have tactile* or *optical sensors.* They can be categorized as follows:

- *Metrology tools* include basic devices such as rulers, protractors, height gages, micrometers, calipers, or optical comparators as well as complex instruments such as coordinate measurement machines (CMMs). Figure 9.8 presents a 5-axis Gantry- style automatic CMM located at the RMU Learning Factory. Also featured in the figure is its calibration tool as well as its tactile probe.
- *3D scanners/cameras* including *2D scanners.*
- *Medical imaging technology* such as *computerized tomography (CT)* is employed in flaw detection and failure analysis in industrial objects as well as assembly analysis.
- Other machines such as *CNCs can be retrofitted with additional equipment* to convert them to CMM or 3D scanners. In addition, the sensors used in game technology (like Kinect) or self-driving cars may also be counted in this category.

Fig. 9.8 5-axis Mitutoyo
Bright Apex 504—automatic
CMM

This chapter will not focus on the metrology tools other than CMMs. Other technologies will be covered in detail in the following sections. Following scanning technologies and associated applications are covered in the following sections:

- 2D Digital Scanners and the Photogrammetry.
- 3D Scanning Technology and Process
    - Measuring Arms, Portable CMMs
    - Scanners with Optical Tracking Devices
    - Structured Light Scanners
    - Portable Scanners
    - Laser Scanning Process.

9.2.1   2D Scanners and the Photogrammetry Process

Photogrammetry is based on deriving accurate measurements from still photographs. It involves taking a set of overlapping 2D photos of an object, building, person, or environment, and converting them into a 3D model using a number of computer algorithms. The lead author and his student took advantage of the free photogrammetry apps previously offered by Autodesk including 123D Catch, Remake, and Recap in digitizing physical objects [8–10]. Some of these apps have evolved over the years, and became integrated into other software tools. Figure 9.9 presents the photogrammetry workflow followed by the RMU team and its details. Additional information about the process is given below as well as an example case where the team scanned a 3D printed skull model based on a CAT scan—Fig. 9.10 a and b.

The photogrammetry process presented in Fig. 9.9 is detailed below with its four steps [8, 11] gives multiple tips on improving the outcome of the photogrammetry process:

1. *Taking images of the object* [11]: A smart phone camera with a high enough resolution (18-megapixels + (MP)) can produce acceptable results for an average application with no accuracy requirements. In addition, a *wide-angle camera* works better since it

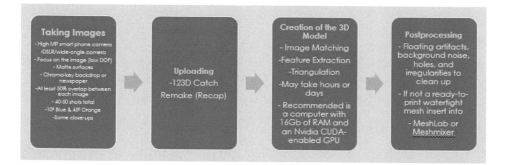

Fig. 9.9 Photogrammetry workflow

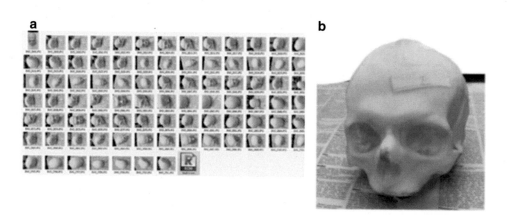

Fig. 9.10 **a** A series of photographs taken. **b** 3D STL model generated by using a smartphone camera and autodesk's remake software [8]

causes least lens distortion. This step involves taking *a series of overlapping images by circling around the object at a low angle looking at the object and a steeper angle to capture the top surfaces* [11].

a. The object needs to be reflective and have matte surfaces to be scanned. Transparent and very shiny surfaces can be covered with a dulling spray or a dry shampoo spray to obtain the desired surface condition.

b. Some of the photogrammetry software tools do not work well with surfaces without features, requiring additional help such as painters' tape or stone effect spray paint.

c. At least a 50% of an overlap between the succeeding images may be needed.

d. 40—50 shots may be enough for one object, taking in two loops (Fig. 9.11)—again at a low angle of $10°$ and a high angle of $45°$ of elevation, also including some close-ups.

e. The object needs to have a big part of the image space by focusing the camera on it and enough color contrast with its background, possibly obtained by a chroma-key backdrop or a newspaper (Fig. 9.10).

f. The object should not be moved during the scanning process. A tripod can be employed to prevent blurry images and consequent elongated scanning process as well as for low-lit environments due to their high exposure time requirements.

g. It can take hours and days to get results from the cloud-based photogrammetry tools based on the size of the scanning and availability of the servers. In addition, any image processing software housed in a local computer should have enough (RAM) memory like 16 GB and good graphics card (GPU).

Fig. 9.11 Recommended photograph orientations [11]

2. *Uploading the images* [11]: Second step of the process is uploading the images into the project library of the photogrammetry software tool by dragging and dropping them. Some software tools have the ability to *check the compatibility of the camera* via its *focal length, principal point, and image sensor format, leading to a number of distortion parameters (bundle adjustment)* before the results can be optimized in the next step. As the pictures are loaded, they may be checked on their impact of the process. A *green* or *red icon indicates* the results of the check, and many rejections may require repeat of step 1 above or editing the images by consistently sharpening them or improving the contrast between the object and background via *a mask called a garbage matte* in a software program like Adobe Photoshop®.

3. *Creation of the 3D model from the images* [11]: Most of the time this step is done automatically, after the completion of the previous step, with some tools presenting options in the form of advanced features for improving the outcome. Following features make up the step 3:

   a. *Image matching:* Most photogrammetry tools convert the images fully automatically into a 3D mesh. However, some tools facilitate manual image matching (*correspondence search*), allowing user to alter the image set before the process starts. During the image matching, the software tool determines which photos are useable for further processing and their overlaps are determined before they can be merged.

   b. *Feature extraction:* While some software tools run through a fully automated workflow, others may permit manual input by the user. This step searches the photos for common features. One method of accomplishing feature extraction is the use of *coded markers*. However, many tools employ the more generic *Structure from Motion* (SfM) method to look for dense textures or patterns on the objects such as texts, wood grain, facial features along with edge points, lines, and corners. Some tools also enrich their data with lighting and shading cues employing a method called, *Shape-from-Shading*. After all the features are identified, they have to go through a process called *Geometric Verification* to eliminate the erroneous features that do not exist or not fully captured by the camera. In this step, the SfM engine creates a transformation that maps feature points between images using a set of algorithms based on *projective geometry*. This step can be observed or interfered with by the users in some tools, and quality of the outcome can be improved by amplifying *keypoint sensitivity and matching ratio*, as well as *changing defaults, and using a different matching algorithm including A-KAZE or brute force method.*

   c. *Triangulation:* Coordinates of the surface points in 3D space are estimated based on the outcome of feature extraction. Lines of sight from the camera to the object in concern are rebuilt in the form of the *ray cloud*. Then, the intersection of the numerous rays determines the 3D coordinates of the object, in the form of a *sparse point cloud*. The photogrammetry software reviews the lighting and texture of the scene to create a *depth map*, advanced tools evening out the lighting across the scanned surface to obtain homogeneous lighting by *delighting*, also including *reverse cal-culation* to cancel out ambient occlusion effects. Even though realistically lit scan models are more desirable for digital visualization, a homogeneously lit model is better suited for full-color 3D printing processes. The dense reconstruction involving the depth map, and the sparse reconstruction that mapped out all features detected earlier are brought into *a 3D mesh format* like FBX, OBJ, PLY, or STL. This stage takes place automatically. However, the user can increase the quality of the result by optimizing settings for *Track Length, Nr. of Neighboring Cameras, and Maximum Points*. Some photogrammetry programs also let the user determine the number of triangles present in the 3D mesh. Carefully reducing the number of triangles in a 3D mesh influences file size and ease of post-processing. Machine learning algorithms can also detect a variety of background objects including buildings and vehicles and filter out moving ones like birds and pedestrians. *Foreground silhouettes, reflectance,* and *irradiance* can be used in enhancing shape data and thin features like power lines can be recreated automatically by *catenary curve fitting algorithms*.

4. Post-processing [11]: After the step 3 is completed leading to the 3D model, the user needs to deal with the issues arisen during the photogrammetry process such as *floating data* not attached to the model, *background noise, holes on the model* along with *other surface irregularities*. In addition, the model may require to be reoriented and rescaled if it is not done by the photogrammetry software. Some photogrammetry tools have the post-processing tools for *patching the holes, smoothing the surfaces by removing irregularities, erasing noise, and remeshing*. Software tools like Meshlab [12] and Autodesk Meshmixer [13] are freely available and can be used from file conversions to a variety of post-processing operations. When the 3D model is completed by being defect free and water-tight, its geometry can be saved as in.STL file format.

9.2.2   3D Scanners

9.2.2.1   How 3D Scanners Work
There are two main categories of 3D scanners based on the way they capture data:

- White-light and structured-light devices that take *single snapshots/scans*.
- Scan arms and portable handheld scanners that *capture multiple images in a continuous fashion*.

The outcome of these scans can be used in i.e. 3D printing/additive manufacturing, animations/simulations, or First Article Inspection (FAI) for quality control. FAI refers to comparing a sample production part with the specifications against which it was designed including the dimensions and tolerances [14]. Software tools like Discus [14] or former Geomagic Qualify are intended for these purposes. The scanned data is presented as a freeform geometry such as a point cloud, unstructured 3D data, a triangular polygon mesh, and can be saved into a variety of intermediate file formats including STL, PLY, or OBJ.

Some 3D scanners can also acquire *surface texture, color and unit information* where they are needed. After the scanning process, different images or scans of the object in concern are brought into a common environment to be *merged into a single and complete 3D model*, as the 3D scanning software aligns (or *registers*) *redundant parts of the images during dynamic referencing* and the duplicate data is eliminated for generating a single surface model. This process can also be executed as a post-processing step. Additional features like *patching holes, smoothing surface*s are applied before a water-tight. STL can be generated.

9.2.2.2   Measuring Arms, Portable CMMs

CMM is a complex device that *measures dimensions/geometry of physical objects by finding discrete points on them.* As mentioned earlier, they can be manual or automatic (repeating the program generated during the first measurement for measuring multiple copies of the same part), and be equipped with *tactile* (mechanical probes—fixed or touch triggered) as well as *optical* (laser or white light). This subsection focuses on one type of portable CMM, the measuring arm.

Resembling articulated robots, *measuring arms* are a type of small *(portable) CMM* equipped with mechanical encoders in their joints. These arms can be equipped with tactile probes and 3D laser scanning heads similar to CMMs, making it possible to integrate scanning and probing in the same effort. On the other hand, these portable instruments need to be fixed on a surface and use their physical link (arm) driven by the user as their positioning method. Basically, the probe's tip can be likened to the tool center point (TCP) of an articulated robot, thus the location of the probe can be determined via forward kinematics as in the case of a tactile probe [15]. Their structure makes them prone to vibrations and other environmental constraints that can affect the quality of the outcome. They also lack flexibility in terms of the locations where they can be employed and the shape and topology of the objects they can scan including small and intricate details (Fig. 9.12).

9.2.2.3   Scanners with Optical Tracking Devices

This type of scanners can track various tools including the positioning of a 3D scanner like Creaform Metra SCAN 3D [15]. They employ an external optical tracking device along with active and passive targets establishing a link between the tracking device and the scanner. This approach adds to the cost of the process. Extending the scanning parameters also adds complexity to the process which can introduce additional uncertainty into the measurements. The optical link between the tracking device and the scanner is considered

Fig. 9.12 FARO Platinum
Measuring Arm located at the
RMU learning factory

Fig. 9.13 Creaform
MetraSCAN 3D [15]

as both the strength and limitation of this technology at the same time. As this type
of scanners present very good accuracy and excellent precision throughout their work
volume, the tracking device must always have a clear and direct line of sight to the
scanner, and most likely have a limited work volume (Fig. 9.13).

9.2.2.4 Structured Light

This type of scanners projects a (stripe) pattern of light (Fig. 9.14a) using a projector or, a
scanned or diffracted laser beam onto an object, and processes how the pattern is distorted

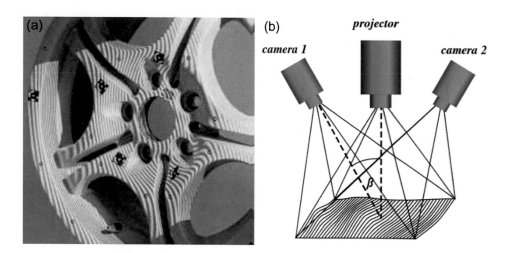

Fig. 9.14 a Structured light projection onto a car wheel [15]. **b** Fringe pattern recognition with two cameras [16]

when light hits that object (Fig. 9.14b) [15]. Two methods of generation of structured light are explained below:

- Laser interference of two wide planar beam fronts leads to regular and equidistant line patterns [16]. These patterns will have a fine resolution in unlimited depth of field (DOF), and can be modified by changing the angle between the beams. Disadvantages include high costs, speckle noise and self-interference with the beams being reflected from the scanned object, having no control over the individual stripes.
- Patterns can also be generated by passing the light through a spatial light modulator including the projection technologies such as transmissive *liquid crystal displays (LCD)*, *reflective liquid crystal on silicon (LCOS) or digital light processing (DLP)* [16]. Even though it can offer multiple advantages, the projection method has some issues, i.e. it may result in discontinuities due to the pixel boundaries of the displays. Some of these can be evened out with defocus.

One or multiple cameras capture the projected pattern. The positioning process between two different pictures captured to perform registration is usually performed off-line based on targets or natural features. If only a single camera is utilized, the relative position of the projector to the camera must be determined before the scanning process. If two cameras are employed, the stereoscopic pair must be calibrated also in advance [15].

High-end structured light scanning process outputs very high-quality data with excellent resolution allowing very small features to be recognized. White-light scanners used in this process can acquire large quantities of data in a single scan. However, the overall project lead time is not reduced due to multiple scans being required to cover different angles on complex parts [15].

9.2.2.5 Portable Scanners

A large number of portable 3D scanners are available to engineers and other practitioners today. These scanners are mainly based on laser or white-light measurements [15]. Laser scanners project one or many laser lines on the object while white-light devices employ a light and shade pattern as explained in the previous section. Both of these technologies process the returned light to extract the 3D data. Hand-held scanners rely on two cameras, enabling them to determine the scanner position with respect to specific points, including positioning targets, the object's natural features or textures (Fig. 9.15a, b). Some of the portable scanners use a mix of those positioning types called hybrid positioning [15, 16]. Advantages and disadvantages of portable scanners are summarized below [15]:

- Advantages: Portable scanners can be transported with easy and are user friendly when compared to other scanner types. They can combine multiple positioning means, including positioning targets (for accuracy), and object's features and texture (for flexibility). They do not require mechanical links found in measuring arms nor optical ones found in optical tracking devices. Having no requirements of links allow them to reach into fine and intricate details, and work in narrow and enclosed areas. They also can scan quickly and acquire a large number of points in a second as they build the 3D mesh live during the scan process [15].

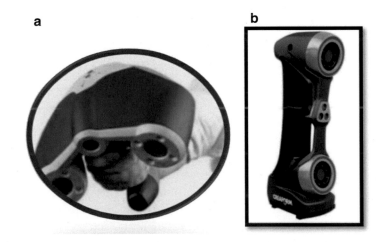

Fig. 9.15 Portable scanners. **a** Creaform Handyscan. **b** Einscan HD, both available at the RMU Learning Factory

- Disadvantages: These types of scanners employ self-positioning on a more local area basis leading to growing errors along with growing scan volume. It is possible to remedy this issue, by employing positioning targets and even photogrammetry. However, these additional efforts increase the set-up time while limiting the possible scan areas' or objects size [15].

All3dp.com is offering a comprehensive look at commercially available portable scanners [17].

9.2.2.6   Laser 3D Scanning

The laser 3D scanner emits a beam of infrared (IR) laser light onto a rotating mirror that effectively paints the surrounding environment with the light including the object [18]. While some scanners' head may rotate, sweeping the object, other scanners are fixed and the object has to be manually or automatically rotated. The object in concern returns the laser beam projected on it, helping establish the 3D geometry in the data acquisition software. These types of scanners also capture measurements on the vertical and horizontal planes, presenting a greater picture of the environment. A laser scanner typically captures data through two kinds of systems [18]:

- *Time-of-flight systems:* Also known as *a pulse measurement system*, this technology works by emitting a single pulse of laser light and determining the distance to the end point by measuring the time it takes for the light to be reflected back to a sensor on the scanner. Light Detection and Ranging (LIDAR) systems are time-of-flight based systems.
- *Phase-shift systems*: This system also uses an emitted laser light, modulated with specific wave forms. The intensity patterns of the laser are modified by the impact by the scanned object's surface. The change between the transmitted and the received laser provides a precise distance calculation. Scanners using phase-shift systems are accurate, fast and provide high-resolution data.

In Practice

Let's say we wanted to 3D scan a person using Structure Core Sensor [19] or a similar scanner.

First, we would set that person within the confines of a *bounding box*, then we would scan them with a scanner. As shown in the figure below, a bounding box (Left) is a measured distance that encompasses the scanned object. The size of the bounding box will determine how large of an object will be scanned.

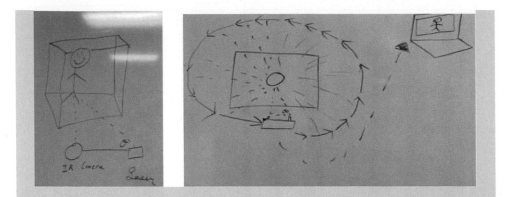

Please follow the Structure Core Sensor instructions given in digital book resources to scan a physical object with no intricate details. You can place the object on to a turntable or a stool with a rotating seat to complete your scan to avoid walking around it as shown in the figure above (Right). Please also follow the data manipulation information given in the digital book resources to complete the process and save the image as .STL file.

9.2.2.7 3D Scanning Workflow for a Laser Scanner Case

This subsection focuses on the 3D scanning workflow (Fig. 9.16) for a case involving the Konica/Minolta Vivid 910 laser scanner as well as the Geomagic Studio software. The lead author followed a similar workflow utilizing a different device, the FARO Platinum Arm scanner with a laser attachment. Details of the workflow is given below in Table 9.2 [20].

Chapter 3 of this book covers the formats used in 3D printing in the presentation of pre-processing for 3D printing. Often those formats like PLY, OBJ, and STL are obtained by 3D scanning.

9.3   Intellectual Property Laws and Ethics

Intellectual property (IP) laws have three folds: including *copyrights, trademarks*, and *patents*. This subsection covers subjects relevant to IP with definitions, examples, and possible applications. Especially copyrights and trademarks may have an important role in reverse engineering efforts of engineers.

Fig. 9.16 3D scanning workflow

Table 9.2 Details of 3D scanning workflow for a Konica/Minolta Vivid 910 scanner/Geomagic Studio software [20]

Phase	Details
Point cloud phase	*Point (cloud) phase* is the step where the scanner sends a collection of points in 3D space to the interfacing computer software. Geomagic studio can filter and reduce the noise (left) (unwanted data points), fill holes, register (aligns) multiple scans based on their redundant areas and merge them
Polygon phase	*Polygon (triangular mesh—left) phase* involves a variety of operations in manipulating the scan data including cleaning, filing holes (right), boundary editing, relaxing boundary, defeature, decimating polygons, sandpaper, relaxing polygons, sharpening, and manifold operations. The digitized model is smoothened or sharpened, additional cleaning is done, and its holes are patched. The number of polygons is also reduced without loosing details
Surface phase	Shape (surface) phase is the last step NURBS and STL (red) models can be generated, allowing 3D printing or transfer to other CAD programs

Copyright is a form of IP applicable to protection of *"original works of authorship including literary, dramatic, musical, and artistic works such as poetry, novels, movies, songs, computer software, architecture"* [21]. It is issued by the U.S. Copyright Office [22]. Multiple resources define key elements or requirements of copyright as presented below:

- The five key concepts of copyright are: *work* itself, *ownership* of the work, what constitutes an *infringement*, and the *exceptions* (to educational institutions, libraries, archives and museum), as well as the *balance* between inventors and authors [23].
- The three requirements for copyright are: *originality, creativity, and fixation*. These are the basic elements that a work must possess in order to be protected by copyright laws in the U.S. as defined below [24]:
  - Originality: A work must be the original work of the author. It cannot be a copy of something else.
  - Creativity: The U.S. Supreme Court has said that a work needs to have a *"modicum (small quantity)"* of creativity to be creative enough for copyright.
  - Fixation: For works to have copyright, they cannot be purely *ephemeral* (lasting for a short time). You cannot get a copyright for a speech, but the moment that is recorded or fixed, it is eligible.

Duration of the copyright is, if it was obtained after 1977, is the remainder of the owner's life plus an additional 70 years. In the case of a copyright obtained for an anonymous publication or one that was published under a false name, it last between 95 and 120 years. © symbol represents a copyright of an original work, in the case of the book, To Kill A Mockingbird which was copyrighted in 1960 [25].

A Trademark is *"a symbol, word, or words legally registered or established by use as representing a company or product"* [26]. It is granted by the United States Patent and Trademark Office (USPTO). Examples of famous trademarks include Apple™, Coca-Cola™, and Nike™. These trademarks are harder to be taken advantage of compared to the unknown trademarks. Each trademarked item is marked with a ™ or a ®symbol, and can be renewed every 10 years with a possibly of lasting forever, as long as it is being used and re-registered [27].

A patent is about "a government authority or license conferring a right or title for a set period, especially the sole right to exclude others from making, using, or selling an invention" [26]. It is also granted by the USPTO, and has 4 different types. Summary of their coverage and durations are given below in the Table 9.3.

Role of IP protections to discourage competitors gaining financial advantage from one's IP. However, it does not guarantee any financial success for the owner of the patent, or the copyright. That lies in the hands of the IP holder and their ability to generate business success. According to npd-solutions [29], if there is a merit in an idea, most likely a competitor will negotiate a license for using the idea or claim that is not novel

Table 9.3 Patents types, their coverage, and time durations (from when it was filed) (not published and made public knowledge) [28]

Patent type	Coverage for/example	Time duration (years)
Provisional*	A protective provision that lasts for 12 months before a proper and formal patent is applied for Can be of any type given below	1 year
Design	The unique visual qualities of a manufactured item including distinct configuration, distinct surface ornamentation, or both Star wars' R2D2 robot	If applied after May 13, 2015, the patent will last for 15 years If applied before May 13, 2015, the patent will last for 14 years from when it was issued
Utility	The creation of a new process, machine, way of manufacturing, or composition of material Blue-ray disc manufacturing process	20 years
Plant	The new varieties of plants, excluding sexual and tuber propagated plants African violet	20 years

by challenging it. A simple (subtle) change or addition like a brake for a scooter that has been patented may prove to be successful.

In the case of Lexmark printers, previous court cases revealed that the reverse engineering for achieving interoperability with a 3rd party created software, is legal and ethical [29]. The software in concern was used in 3rd party print cartridges made by SCC as Lexmark tried to prevent SCC from selling its cartridges rather than the original ones it markets. Additional IP cases may be covered through the question section of this chapter.

9.4 Non-industrial Use Cases

Non-industrial use cases encompass the virtual reconstruction covered in Chap. 7 including efforts in preserving and replicating historical artifacts, possibly identifying primate or dinosaur bones, or crime victims. Biomodelling and custom fabrication of implants also occupy a sizable share in non-industrial uses. This section covers the biomodelling process via medical imaging leading to custom prosthetic implants. This is different from fabrication of custom orthotics, also covered in Chap. 7, which rely on external scanning means or simply 3D scanners.

Biomodelling is basically the result of digitization of human or animal anatomy, its tissues, soft or hard, and its organs. It can be considered reverse engineering the anatomy of the living being including its issues—tears/fracture, or tumors. It can also be handled

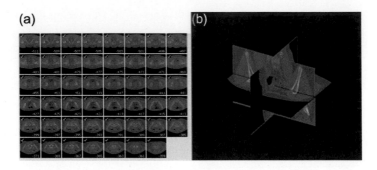

Fig. 9.17 **a** Medical images of a human pelvis in a sequential order. **b** Hip joint cross-section in three orthogonal planes [30]

at microscopic levels. Biomodels can be employed in *pre-surgical planning, educational and training purposes* as well as *custom-prosthesis and implant design* [30]. The work covered in this subsection was done with Materialise Mimics software by the lead author. Today, there is a free alternative to this software, 3D Slicer, an open-source multi-platform software tool [31]. Workflow for biomodelling is given as follows [30]:

- A series of sequential MRI (Magnetic Resonance Imaging) or CAT (Computerized Assisted Tomography) scan images are required before a 3D rendering of a human anatomy element is generated—Fig. 9.17a.
  - A minimum of at least 30 slices with a slice depth of 1 to 4 mm and a medium capture resolution are needed to generate 3D bone, muscle, or organ models with clarity.
  - Before the 3D rendering process takes place the slices must be put into sequence in the orientation plane in question.
  - Once each set of slices are sorted, three cross-sections shown previously Fig. 9.17b are produced allowing the slices to be cycled from the *anterior* to *posterior*, *top* to *bottom*.
- The users can select a certain type of tissue, hard or soft, by applying visual filters—Fig. 9.18a, b.
  - These filters can identify the density of the tissue based on its color or shades.
  - Each filter can be readjusted by changing its threshold values through a histogram tool.
  - This custom threshold feature can then be used for identification of anomalies including tumors as well.
- The 3D model is moved to Geomagic Studio to be cleaned of noise and made water-tight if there are voids—Fig. 9.19.
- Once the 3D computer model is ready, it can be saved in.STL format to be printed or imported into the Materialise 3-Matic software for developing custom bone implants.

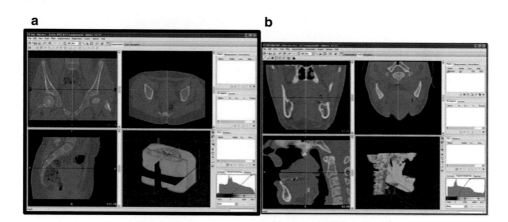

Fig. 9.18 **a** Different slice orientations and the histogram tool. **b** Resulting 3D bone models [30]

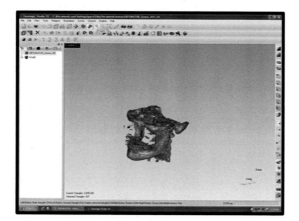

Fig. 9.19 Mimics model imported into Geomagic Studio [30]

- Stratasys Dimension ELITE was used for 3D printing of the physical models—Fig. 9.20.
- Following items define the workflow for designing a custom fit prosthetic (cranioplastic) bone once the needed biomodel is completed in Mimics software [30].
  - The Mimics 3D model is moved into the 3-Matic program—Fig. 9.21.
  - The designers switch to the sketch mode to complete bone by drawing its wall structure—Fig. 9.22.
  - Then the outline of the void is marked by the designers—Fig. 9.23.
  - After the bone implant geometry is generated—Fig. 9.24, a physical bone implant can be printed and applied via surgery to the patient—Fig. 9.25.

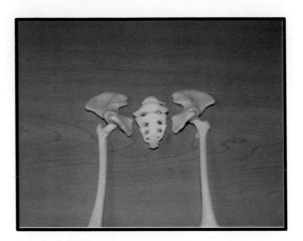

Fig. 9.20 3D printed physical hip model segments [30]

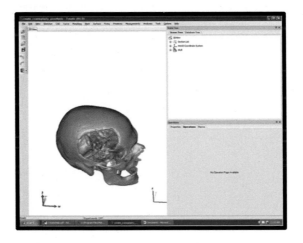

Fig. 9.21 Mimics 3D model input to the 3-Matic software [30]

Review Questions

1. Define reverse engineering? List the two main components of reverse engineering?
2. What is reverse engineering technology? Elaborate on the purpose of the hardware and software tools employed.
3. Define the photogrammetry process and role of 2D scanners.
4. List the different classifications of 3D scanners mentioned in this chapter.
5. List the two major categories of scanners based on the way they capture data.
6. Define the laser scanning process.

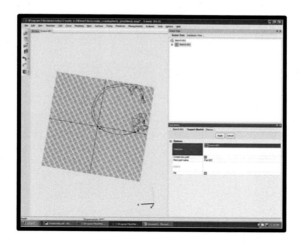

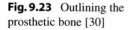

Fig. 9.22   Sketching the prosthetic bones walls [30]

Fig. 9.23   Outlining the prosthetic bone [30]

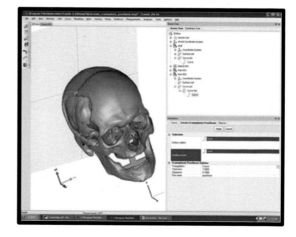

7. What is the difference between time of flight and phase shift laser scanner?
8. Measuring arms are considered as portable CMMs, but need to be fixed at a location before use. How do they measure the position of point in 3D space if these arms are using a tactile probe?
9. What is the difference between laser-line- and white light-based portable scanners?
10. What is the triangulation process used in 3D scanners?
11. What are the advantages and disadvantages of portable 3D scanners compared to other types of scanners?
12. What are the steps of the 3D scanning workflow using a laser scanner? Explain each step concisely.
13. How do structured light scanners work? Explain it concisely.

Fig. 9.24 Resulting
cranioplastic bone implant [30]

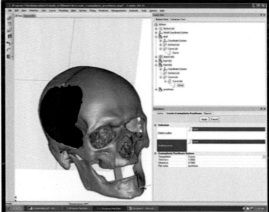

Fig. 9.25 Applying the
implant after both the skull and
the implant was 3D printed
separately [30]

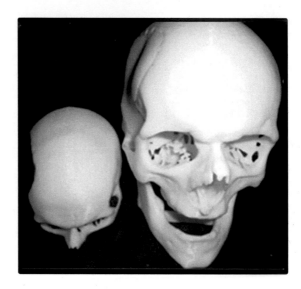

14. What are the formats generated by the Structure Core Sensor?
15. What are the disadvantages of the.STL format which is the most common intermediate file format in 3D printing applications?
16. What are the three forms of IP laws?
17. Define the term "copyright" and give an example.
18. What are the five key concepts of a copyright? Explain each briefly.
19. What are the three requirements of a copyright? Explain each briefly.
20. Define the term "trademark" and give an example.
21. Define the term "patent" and give an example.

22. What are the four patent types? Explain each one briefly, in terms of their coverage, durations, and give one example for each.
23. What is the workflow for the biomodelling process? Explain each step briefly.
24. What is the workflow for making custom cranioplastic implants? Explain each step briefly.

Research Questions

1. What is the "dirty room" concept pertaining to reverse engineering an existing software?
2. Find, review, and summarize a reverse engineering methodology case study in a PowerPoint presentation in 5 slides or less.
3. Disassemble a children's toy to prepare a glass box model, after preparing a block box for it.
4. Find a CMM program for measuring simple geometric features like a hole, its center, and diameter.
5. Find and use a smart device photogrammetry app and an associated cloud base software to generate a 3D CAD model from 2D digital photographs.
6. Find a video on optical trackers, watch it, and summarize the inner workings of these types of scanners.
7. Which of the intermediate file formats (? OBJ, .PLY, .VRML, .AMF, and .3MF) is becoming more common in 3D printing in route to replacing the .STL format? Research the subject and discuss your findings.
8. Availability of free form curves and associated surfaces improve the results of a 3D scanning process since they can help better represent curved edges and surfaces. Research and define free-form curves available to engineers with one example for each type.
9. What is the application area of .VRML format?

Discussion Questions

1. Prepare a customer product review form for a new tool designed for handicapped persons after discussing with your peers (Hint: Use Kinzel's reverse engineering materials as an example).
2. Find, review, and summarize a reverse engineering methodology case study where the process did not lead to infringement.
3. Suggest improvement(s) for the toy in concern given in question 3 of research questions.
4. Discuss a case where ergonomics and/or human factors was the main reason for reengineering of the product in concern after its reverse engineering.

5. Review a copyright infringement as it applies to an interoperability of case of software, and summarize in a page.
6. Review a trademark infringement case and summarize in a page.
7. Review a patent infringement case and summarize in a page.

References

 1. Dym, C. L., Little, P. (2000). *Engineering Design: A- Project-Based Introduction*. New York, NY: John Wiley.
 2. Noorani, R. (2006), *Rapid Prototyping Principles and Applications*. Hoboken, NJ: John Wiley and Sons Inc.
 3. Kinzel, G., Piper, J., Murdell, R., Pham, P., Detrick, M. (2000), Reverse Engineering Module Supplement for the NSF sponsored Gateway Coalition (grant EEC-9109794).
 4. Chinese wall—Wikipedia, located at: https://en.wikipedia.org/wiki/Chinese_wall, accessed September 16, 2022.
 5. Schwartz, M., How to: Reverse Engineering (November 12, 2001), https://www.computerworld.com/article/2585652/reverse-engineering.html
 6. U.S. Patent and Trademark Office, located at: https://www.uspto.gov, accessed September 16, 2022.
 7. Google Patent Search, located at: https://www.google.com/patents, accessed September 16, 2022.
 8. Walter, D.J., Sirinterlikci, A. (2017). Utilization of Freeware and Low-Cost Tools in a Rapid Prototyping and Revers, American Society for Engineering Education (ASEE) Conference and Exhibition, Columbus, OH.
 9. Autodesk Photogrammetry Review 123d Catch, located at: https://3dscanexpert.com/autodesk-photogrammetry-review-123d-catch/, accessed September 16, 2022.
10. Autodesk Remake Review, located at: https://3dscanexpert.com/autodesk-remake-review/, accessed September 16, 2022.
11. Photogrammetry guide and software comparison, located at: https://formlabs.com/blog/photogrammetry-guide-and-software-comparison/, accessed September 16, 2022.
12. MeshLab, located at: https://www.meshlab.net/, accessed September 16, 2022.
13. Autodesk Meshmixer, located at: https://www.meshmixer.com/, accessed September 16, 2022.
14. First Article Inspection Software, located at: https://discussoftware.com/, accessed September 16, 2022.
15. Allard., P-H., Lavoie, J-A. (2014). Differentiation of 3D scanners and their positioning method when applied to pipeline integrity, 11th European Conference on Non-Destructive Testing (ECNDT 2014), Prague, Czech Republic.
16. Structure light 3D scanner, located at: https://en.wikipedia.org/wiki/Structured-light_3D_scanner/, accessed September 16, 2022.
17. Best 3d scanner diy handheld app software, located at: https://all3dp.com/1/best-3d-scanner-diy-handheld-app-software/, best-3d-scanner-diy-handheld-app-software.
18. Understanding laser scanners, located at: https://www.faro.com/en/Resource-Library/Article/understanding-laser-scanners, accessed September 16, 2022

19. Sirinterlikci, A., Bill, C., Connolly, P, Wolfe, A., Farroux, L. (2020), Introduction to Advanced Manufacturing Techniques Module Supplement for the PA DCED Training Career Program Grant (Empowering PA Manufacturing Workforce Project).

20. Sirinterlikci, A., Uslu, O., Czajkiewicz, Z. (2008), Replicating Historical Artifacts: Robert Morris Case Study, SME 3D Scanning Conference, Lake Buena Vista, FL.

21. What does copyright protect? located at: https://www.copyright.gov/help, located at: https://www.ualberta.ca/2016/11, accessed September 16, 2022.

22. Copyright Office|USAGov, located at: https://www.usa.gov/copyright-office, accessed September 16, 2022.

23. A Brief Overview of Copyright: 5 Key Concepts| The quad, located at: https://www.ualberta.ca/2016/11, accessed September 16, 2022.

24. Copyright Law Basics—Copy Quick Reference Guide, located at: https://guides.library.unt.edu/basics, accessed September 16, 2022.

25. How Long Does Copyright Protection Last? (FAQ), located at: https://www.usa.gov/copyright-office/faq-duration.html/, accessed September 16, 2022.

26. OxfordLanguages, located at: https://oxfordlanguages.oup.com, accessed September 16, 2022.

27. Keeping your registration alive | USPTO, https://www.uspto.gob/trademarks/maintain/keeping-your-registration-alive/, accessed September 16, 2022.

28. Patent process overview—USPTO, https://www.uspto.gov/patents/basics/patent-process-overview/, accessed September 16, 2022.

29. Study, located at: http://ethics.csc.ncsu.edu/intellectual/reverse/study.php/, accessed September 08, 2008.

30. Sirinterlikci, A. (2012). Teaching Biomedical Engineering Design Process and Development Tools to Manufacturing Students", 2012 ASEE Annual (American Society for Engineering Education) Conference and Exposition, San Antonio, TX.

31. 3D Slicer Image Computing Platform, located at: https://slicer.org, September 16, 2022.

Special Topics in Digital Manufacturing 10

The authors feel that special topics in digital manufacturing may include subjects like cost modeling and estimation, sustainability, and environmental, safety, and health (EHS) subjects of 3D printing and additive manufacturing as well as the mixed reality realm with augmented, virtual, and hybrid reality applications, their associated hardware and software. Even though this book touches base on IoT, especially the IIoT, and digital twin concepts in Chap. 8, most of smart manufacturing including automation and control subjects, machine learning and data analytics, as well as robotics are not within the scope of it.

10.1 Cost Modeling and Estimation for 3D Printing and Additive Manufacturing

This subsection has been adapted from the lead author and Ergin Erdem's unpublished work, titled, A Cost Estimation Study of Fused Filament Fabrication (FFF) Parts [1].

While there has been much more effort spent in technical aspects of 3D printing and additive manufacturing, especially in materials, processes, instrumentation and controls, and applications development, there have not been many studies focusing on their financial aspect through cost modeling work. A few works are presented below in Table 10.1.

In addition to the cost modeling work presented in Table 10.1, there have been research activities that focused on cost analysis of various applications relying on 3D printing:

- A study conducted by Dong and the co-authors specifically focused on *cost analysis of 3D printed circuit boards (PCBs)*. The authors argued that the proposed system might be used for significant cost and energy savings and might contribute to the development of the next generation of green technologies. In addition to the cost analysis, the work

© The Author(s), under exclusive license to Springer Nature Switzerland AG 2023
A. Sirinterlikci and Y. Ertekin, *A Comprehensive Approach to Digital Manufacturing*,
Synthesis Lectures on Mechanical Engineering,
https://doi.org/10.1007/978-3-031-25354-6_10

Table 10.1 Noteworthy studies focusing on cost modeling for 3D printing

Reference	Findings/points deduced
Ajay et al. [2]	– The first part of the study quantified the *material consumption* and *electricity use* in the FFF extrusion process—concluding that electricity use took up 32% of the total cost – The second part of the study detailed the energy use and identified the sensitivity of the various factors including *motion energy, heating, and cooling requirements,* especially looking into *motor speed for the extruder, heater, and cooling fan* – Authors claimed that the electricity costs are omitted from the cost figures. However, this cost component can vary from country to country. Thus, generalizing this claim will be difficult
Piili et al. [3]	– The study focused on the cost structure for the DMLS process, including its *build times, material, machine, and energy costs* – The process in concern relies on *high initial investment*, and that was found to be much greater than the material and energy costs
Hitch [4]	– Presents the *five secrets costs of post-processing* for the New Equipment Digest (NED) magazine based on a white-paper prepared by Todd Grimm – *Loss of production time* occurs (as much as 60% i.e. for a case at Newell Rubbermaid) while the newly printed parts are being post-processed keeping operators busy. Thus, additional parts cannot be printed. Every 6 h printing requires on average 1 h of post-processing, increasing the total process time by at least 17% – Secondary and third impact comes from *increased labor costs* since most likely the highly skilled and paid people are being employed in post-processing while taking away their productivity in terms of innovation for new concepts and prints – Fourth concern is the *unreliable quality for removing support materials* due to the simple tools used such as xacto knives and sand paper – The fifth concern is the *environmental impact through use of ultrasonic cleaners filled with sodium hydroxide* for the FDM process, producing large amounts of hazardous waste and subsequent contamination

provides an analysis of the impact on the environment as well, which has proven to cut energy consumption, conserve material, and consequently reduce waste [5].

- Winters and Shepler provided a *cost analysis on making optomechanical equipment for open-space spectroscopy*. They compared the corresponding costs of *3D printing of certain parts of the interferometer or buying those components from the local hardware shop*, and concluded that 3D printing is much cheaper as compared to the commercial purchase option [6]. In a similar fashion, Zhang and the co-authors compared the cost of hardware components for *an optical system* (used for research and teaching purposes) manufactured by *3D printing or traditional methods*. The study revealed cost reductions of up to 99% for some of the components while a reduction of 97% was achieved across the board, making these devices accessible to broader audiences [7].

- Mostafa and the coauthors employed the design of experiments (DOE) approach for the FDM process and Nylon 12 material. The group measured *ultimate tensile strength (UTS), ultimate flexural strength (UFS), modulus of elasticity, print time and volume* based on ISO compliant specimens, and presented *the strength to cost ratio of different printing configurations.* Using the Taguchi's prediction model, they also *optimized the printing configuration with mechanical properties, printing time, and volume prediction* [8].

- Franchetti and Kress compared the costs of 3D printed and injection molded parts. Using cost analysis, they performed calculations on *break-even points based on the lot sizes, part mass, and density of the object.* Their research carries significant importance based on the fact that the *feasibility of using additive manufacturing techniques for the mass production* as compared to the traditional methods was studied in terms of the costs [9].

- Baumers and the coauthors compared the *costs of two different metal 3D printing processes,* EBM and DMLS. The study indicated that both processes have high specific costs which creates a strong barrier for the adoption of these processes in manufacturing applications, however, economies of scale might be still achievable to some extent. The authors also deduced that the *3D printing applications might benefit significantly in terms of cost from the market pull and technology push modes of innovation* [10].

- The lead author and his collaborators studied the cost of silicone rubber molds made via indirect rapid tooling as a low-cost alternative to 3D printed replicas of historical artifacts [11].

- Zhang and Wang investigated the possibility of adopting *3D printing on the meniscus from both supply chain and patient perspectives.* Various data sources were used such as online resources, literature and citations in addition to making certain assumptions for obtaining cost figures. Their analysis not only considered *the cost of traditional and 3D based transplantation from the patients' perspective* but also focused on post transplantation costs resulting from associated risks for such patients [12].

- Fernandez-Vicente and the collaborators focused on *replacing the traditional process of making custom orthotics with combined 3D printing and non-manual finishing.* The cost analysis was conducted to demonstrate the feasibility of 3D printing for making custom orthotic devices [13].

- Researchers, like Smektala and the collaborators, conducted a *cost study for manufacturing low-cost silicone renal phantoms* for surgical training [14]. The authors suggested that using basic 3D printing for fabricating the mold for casting silicone renal replicas is a low-cost alternative, and the model created by this approach provided shape and elasticity compatible with the living organ and has similar mechanical strength.

- Other researchers, like Cameron and Gordon, looked into the *cost analysis of the support structure of 3D printed objects.* They calculated the *cost of an optimal support structure with respect to overall mass and structural integrity* using a recursive algorithm [15]. Their approach was based on *tree-like structure using predetermined optimal grouping patterns.* The developed algorithm is then adapted for analyzing loading of those structures with respect to compressive downward force and maximum force moments.

- Rather than specific case studies seen above, some works provided *a broader approach on economics of 3D printing instead of specific applications*. To cite an instance, Feldmann and Pumpe embarked on a study on *providing comprehensive information about value drivers by developing a framework for investment decisions based on the economic value added*. For assessment of value drivers in a global supply chain, the authors provided an empirical study showcasing companies across different industries [16].

Using the limited literature review presented above, the authors can arrive at the following summary, as they saw efforts on:

- Cost analysis of specific additive manufacturing applications including PCBs, optical components, ceramic tooling, prosthetic and orthotic devices, phantoms for surgical training. Some studies focused on implant/transplant costs as well as post implant/transplant costs for the medical applications.
- Comparison of costs of 3D printing and competing traditional manufacturing processes as well as making of components with 3D printing versus buying them. This type of effort included employment of break-even analysis on the lot size to explore the feasibility of additively manufacturing components in concern for mass manufacturing.
- Comparison of two or more 3D printing processes to be used in process selection.
- Understanding the relationship between the cost and mechanical properties in 3D printed parts, en route to achieving process optimization including support structures.
- A broader approach on economics of 3D printing, providing comprehensive information about value drivers by developing a framework for investment decisions based on the economic value added.
- Impact of market pull and technology push modes of innovation on the cost of 3D printing applications.

10.1.1 A Cost Model for the FFF Process

A study by Lan and Ding developed an elaborate costing model for the SLA process for quoting by service bureaus [17]. This model was adapted by the lead author and employed for FFF printing of Nylon hydraulic filter components with the MakerBot Replicator 2XTM machine [18]. The Makerbot printer was modified to allow Nylon derivatives to be printed including use of blue painter's tape or G-type phenolic glasses on its build platform.

Production cost of the FFF process is the sum of all costs associated with raw printing materials and their disposal, purchased items such as blue painters tape or G-type phenolic glasses/the glue used for modification of the build plate in addition to the process costs including labor cost, material cost, software and equipment costs, and power consumption. All of these cost components can be reorganized based on the stages of the process to analyze the cost of the FFF process. For any 3D printing system, the total cost can be broken into those of the three stages of the process; the pre-processing (C_{PRE}), build (C_{BUILD}), and post-processing (C_{POST}). The total cost C_{TOTAL} is the algebraic sum of the cost of these three stages given in Eq. 10.1.

$$C_{TOTAL} = C_{PRE} + C_{BUILD} + C_{POST} \qquad (10.1)$$

The preparation for the FFF process involves completing everything necessary before the build starts. This step includes the CAD work, conversion of the CAD file into the STL geometry, and generation of the print NC code (in .gcode format) by slicing the STL geometry along with the addition of extrusion paths and settings, and other process settings. Additional tasks such as support generation and material load/unload, leveling of the build plate, and diagnostics check are also considered in pre-processing costs along with fixing of the STL files. The pre-processing cost is defined as C_{PRE};

$$C_{PRE} = (T_{CAD} + T_{STL} + T_{PREB}) \times (C_{OPR} + C_{CS}) + T_{SETP} \times C_{OPR} \qquad (10.2)$$

where T_{CAD} is the time spent in preparing the CAD model of the print; T_{STL} is the time for generating the STL file; T_{PREB} is the time for pre-building preparation including scaling, orientation and slicing of the STL as well as the process parameters planning while T_{SETP} encompasses all printer set-up procedures. C_{OPR} is the labor cost per unit time of pre-build preparation and set-up assuming that the same person is handling both. C_{CS} is the rate for the cost of computer hardware and associated software.

Build cost includes the build time multiplied by cost per unit time for operating the machine and the costs of the modeling and support material. Operating cost per unit time includes the machine depreciation, power use, and operating overheads. Build time can be further divided into printing and idle time. The printing time is the actual time for printing the model and its support. Idle time is non-productive nor value-adding, including the z movement, or waiting time due to the controller reading the code.

$$C_{BUILD} = (T_{PRINT} + T_{IDLE}) \times C_{PRINTING} + C_{MODEL} + C_{SUPPORT} \qquad (10.3)$$

where T_{PRINT} is the printing time; T_{IDLE} is the idle time; $C_{PRINTING}$ is the building cost per unit time; C_{MODEL} is the cost of the modeling material; $C_{SUPPORT}$ is the cost of the support material.

Post-processing costs involve detachment of the part from the machine, removal of the support material, and subsequent finishing operations including infiltration of the parts with resin and sanding (if necessary). The total cost for post-processing can be defined as:

$$C_{POST} = (T_{REMP} + T_{REMS} + T_{FINISH}) \times C_{OPR} + C_{MAT} \qquad (10.4)$$

where T_{REMP} is the time span for removal of the model from the machine; T_{REMS} is the time span for removing the supports from the model; T_{FINISH} is the time spent for infiltration and sanding as finishing operations; C_{OPR} is the cost of labor per unit time; C_{MAT} is the cost of materials associated with the post-processing operations including those of infiltration or sanding.

In-Practice

By following the industrial project example given below, please conduct a similar cost analysis of a 3D printing process.

This example focuses on the cost of hydraulic filter end caps printed using a Nylon derivative by a MakerBot Replicator 2XTM machine [18]. The pre-processing cost is calculated based on Eq. 10.2 (given above in the chapter and also repeated below) and associated data given below:

$$C_{PRE} = (T_{CAD} + T_{STL} + T_{PREB}) \times (C_{OPR} + C_{CS}) + T_{SETP} \times C_{OPR} \qquad (10.2)$$

$T_{CAD} = 45$ min $= 0.75$ h (average CAD time for generation of hydraulic filter parts).
$T_{STL} = 5$ min $= 0.083$ h (includes software start and STL parameter selection times).
$T_{PREB} = 5$ min $= 0.083$ h (includes input of the STL file, setting the process parameters, leveling of the machine).
$T_{SETP} = 10$ min $= 0.167$ h (includes modification of the build platform by application of blue painters' tape or G-type class film and consequent heating of the platform).
$C_{OPR} = \$10/h$ for an RMU research assistant or $\$21.25$ in average for partnering industry technicians.
$C_{CS} = \$800 + \$149 = \$949/6240$ h $= \$0.152/h$ (The cost of a PC, perpetual license cost of Simplify 3D software and 3 years as the lifetime of the PC assuming that it is utilized with 2080 h per year).
$C_{PRE} = (0.75 + 0.083 + 0.083$ h$) \times (\$10/h + \$0.152/h) + 0.167$ h $\times \$10/h = \$10.969/$part when an RMU research assistant is employed.

This cost element C_{PRE} is consistent with any number of prints in a given batch and if there are more parts to be printed, this cost will go down since the figure above assumes that the machine is printing a single part only. Even with twice as costly industrial technician's hourly contribution to this cost element, C_{PRE} will be slightly less when printing parts in a batch more than two parts.

The build costs are calculated based on Eq. 10.3 (given above in the chapter and also repeated below) and associated data given below:

$$C_{BUILD} = (T_{PRINT} + T_{IDLE}) \times C_{PRINTING} + C_{MODEL} + C_{SUPPORT} \quad (10.3)$$

$T_{PRINT} + T_{IDLE} = 0.5$ and 2 h for small or large filter end caps respectively.
$C_{PRINTING} = \$2500/3650$ h or $\$0.685$/h (The printer was purchased for \$2500). The life figure of 3650 h for MakerBot Replicator 2XTM printers was obtained from a study by King [19] + \$0.035 (0.50 KW power consumption for 1 h of print [19] and \$0.07 cents for 1 KWh [20] = \$0.72/h).
$C_{MODEL} + C_{SUPPORT} = \$20/453.6$ g (Assuming \$20 is the average cost of Nylon derivatives per 453.6 g) $\times$ (7–43 g) = \$0.309–1.896/part for small or large filter end caps respectively.
$C_{BUILD} = 0.5$ h $\times$ 0.72/h + \$0.309 = \$0.669/part—for small filter end caps.
$C_{BUILD} = 2$ h $\times$ 0.72/h + \$1.896 = \$3.336/part—for large filter end caps.
C_{BUILD} varied between \$0.669/part to \$3.336/part depending the size of the end caps. These figures can be further increased if a 10% failure rate is added to the mix, leaving us with slightly higher numbers. These figures are also for a single part batch.

The post-processing costs are calculated below based on Eq. 10.4 (given above in the chapter and also repeated below) and associated data given below:

$$C_{POST} = (T_{REMP} + T_{REMS} + T_{FINISH}) \times C_{OPR} + C_{MAT} \quad (10.4)$$

$T_{REMP} + T_{REMS} + T_{FINISH} = 10$ min = 0.167 h and $COPR = \$10$/h or \$21.25 (as mentioned above if the industrial partner's technician is involved)
  = 0.167 h * 10 = \$1.67/part when an RMU research assistant conducted the work.

Material Cost for post processing (C_{MAT}) is assumed to be zero, since the support material was removed by breaking it away. At the end of this project, the industrial partner was experimenting with infiltration and coating of the printed parts with other materials to improve end cap performance. Since this effort was not conclusive at the time, it was not included in the costs.

The total cost of a MakerBot printed parts could be as little as $10.969 + 0.669 + 1.67 = $13.308 for the small end caps or as much as $10.969 + 3.336 + 1.67 = $15.975 for the large end caps. In addition, a $30 dehumidifier with a 0.600 KW power requirement is used for 2 h for removing moisture from multiple spools before the printing operation. Drying process' addition to the total cost will be 1.2 KWh * $0.07 = $0.084/spool or $0.000185/gram if a single spool is placed into the machine. Considering the range of the parts in terms of their weight, this process does not add a significant amount to the total cost.

The example given in the In Practice section above does not reflect the impact of inflation. Thus, the following additions were made considering changes in labor, material, and energy costs for a multi-year span, at the time this study was completed [1]:

- It is worth considering the effects of the increasing pre-processing costs associated with the labor. According to the Bureau of Labor Statistics [21], the labor cost index as of December 2005 was 100 for universities, and this index was raised to 126.7 in December 2015. This change indicates an average annual percentage increase of 2.395%. Assuming that the same trend continues, the associated cost per hour would be $10.00/h in 2016 (year 0), $10.24/h for 2017 (year 1), and $10.48/h for 2018 (year 2) in a three-year span. A second assumption is made on the cost of computer hardware and software and this cost is not considered to be affected by inflation. One of the reasons behind this assumption is that there are three scenarios that are considered in this analysis. For the year 1 and year 2 of the utilization cycle, because of the fact that computer hardware and software costs are paid upfront, we assume that these costs are not affected by inflation. Following similar line of reasoning above, the cost of pre-processing per unit can be calculated as $10.969, $11.229, $11.494 in years 0, 1, and 2.
- As similar to the increasing cost of labor, it would be worthwhile to consider the effects of the increasing material and energy costs. Since the MakerBot Replicator 2XTM printers have been purchased upfront, as in the case of computer and software, the inflationary effects on the printer is not taken into the consideration. However, the energy and material costs are subject to the inflationary effects. In order to calculate the inflation rate, the 10-year geometric average is considered. As such the *Consumer Price Index* (i.e., CPI) is considered for calculating the corresponding increases in materials and energy. According to the U.S. Inflation Calculator [22], the CPI index of 196.8 at the end of December 2005, becoming 236.525 at the end of December 2015, leading to an average inflation rate per year of 1.856%. As such the cost of electricity and material can be calculated similarly. This leaves us with a minimum unit processing cost of $0.669 in 2016 (year 0), $0.681 for 2017 (year 1), and $0.694 for 2018 (year 2). As such the maximum unit varying cost ranges between $3.336/unit in year 0, $3.398/unit in year 1, and $3.461/unit in year 2 respectively.

Table 10.2 Minimum and maximum total cost per unit with respect to years [1]

Year	Minimum cost ($)	Maximum cost ($)
2016	13.31	15.98
2017	13.61	16.31
2018	13.93	16.65

- Finally, the post processing cost would be $1.67/part in 2016, $1.710/part for 2017 (year 1), and $1.751/part for 2018 (year 2) following the same pattern used early in the inflation analysis.
- The associated minimum (small end caps) and maximum (large end caps) total costs based on the corresponding years are provided in Table 10.2.

The effect of the inflation was considered for different scenarios [1]. The first scenario is using the printer for 2 years, approximately 5 h/day printing which indicates a higher duty cycle, its value being approximately 20.83%. The second scenario is utilizing the machine for 3 years, with a moderate duty cycle having the value of 13.88%. The third case can be centered on using the machine for 5 years, which corresponds to using the printer approximately 2 h/day, which corresponds to a light duty cycle with a value of approximately 8.33%. Since the computer and machine life is determined to be 3 years, the last option was not considered [1]. Additional information on this analysis is available via the lead author of this book.

10.1.2 Building the Case for Business

GE Additive is one of the pioneering and leading companies in the industrial 3D printing and additive ecosystem. According to a playbook prepared by the company [23] "manufacturers must look at how additive (manufacturing) impact the whole system—from part costs to product performance improvements to supply chain impact to new revenue systems." Thus, just the cost analysis alone might hinder companies' ability to gain larger *returns on investment (ROI)* as well as new business opportunities. The playbook sketches out a four-step process for creating an AM business case [23]:

- Building a cost model—How can AM lower part costs? In traditional manufacturing increasing design complexity results in higher costs while AM costs remain constant with changes in the design, thus AM can easily facilitate *design freedom*. With this first step, companies need to evaluate possible parts for AM, using factors like complex geometries, high-labor parts, fabricated assemblies and possibility of part consolidation, durability and weight improvements, performance improvements, obsolete parts, customization, and low-volume parts [23].
- Evaluating performance factors—How will AM impact PLM and its costs? Companies need to evaluate parts with large ROI potential, considering multiple positive outcomes. The playbook lists a few factors which can be applied to a variety of products: design

Table 10.3 Supply chain simplification with AM—GE aviation mid-frame superstructure [23]

Parameter	Conventional manufacturing	Additive manufacturing
Parts	300	1
Engineers involved	60	6–8
Manufacturing sources	50+	1
Data systems	40	1
Repair sources	5	1

freedom, enhanced performance, enhanced reliability and reduced risks, weight reduction, improved efficiencies, durability, less defects during manufacture, and improved sustainability [23].

- Identifying supply chain disruption—How will AM streamline the manufacturing process and resolve its supply chain issues? AM can help manufacturers streamline their supply chain and optimize the process. The playbook presents the average cost of *purchase order (PO)*–$217 average across industries (low PO costs are $59 while high costs in some industries can run $741 on average). PO costs increase for each additional supplier needed to complete a product, also causing increases in labor costs, manufacturing sources, inspection and repair sources. Following list of factors may impact the supply chain of a manufacturing enterprise: serial production of mixed designs and sizes, part consolidation, inventory reduction, waste reduction, purchase order reduction and freight savings, streamlined supplier base, in-house tooling operations, reduced workflow, lead-time reduction, maintenance/repair and overhaul improvements, single source manufacturing. Table 10.3 illustrates a case where AM simplified the supply chain for a GE Aviation frame structure [23].

- Determining the ROI—Employing company's business goals and ROI analysis, can the company justify applying AM into its manufacturing processes? Assembling the critical information/results from the three previous steps, cost models and subsequent analysis, performance factors, and supply chain optimization, a company can determine the ROI and identify the parts with the greatest potentials for AM. The playbook presents a checklist, which is a subject of an end-of-chapter question [23].

10.2 Sustainability and Environmental Issues for 3D Printing and Additive Manufacturing

Even though it has been changing, in the recent past, the technology development took precedence over its safety, health, or environmental impact, also including sustainability. Early PCs and their components such as PCBs filling landfills, industrial chemicals leaching into our water streams, and plastic bottles and bags harming sea life are some of the major concerns resulting from our engineering successes without thinking about their end-of-life impact to our environment. The lead author of this book started investigating the impact of nanomanufacturing and 3D printing in the mid-2000s on safety and health of the workers and engineers involved, producing a few scholarly works in the process [24–26]. His collaborative work with an RMU chemist (Paul Badger) and an RMU environmental scientist (Daniel Short) on expanding his original EHS issues in rapid prototyping paper from the RAPID/3D Imaging Conference [24] yielded a Rapid Prototyping Journal article [25]. This paper was followed by another work focusing on the contents of the early epoxy-resins (a form of photopolymers) used in the SLA process, identifying antimony, a carcinogen, and its potential impact to the environment [26]. The summary of the study is given below:

- Early photopolymers used for SLA 3D printing were one of several polymeric materials on the consumer goods market containing large amounts of antimony. These materials utilize antimony as a photo-initiator in the polymerization process. Antimony is a toxic heavy metal whose environmental concentration and atmospheric flux has recently been found to be increasing. With the projected growth in the 3D printing industry, the health and safety aspects associated with antimony containing photopolymers and their safe disposal of 3D printed models were a concern. The antimony content of three different types of polymeric materials (photopolymers, polyethylene, and brake pads) was compared. Leaching tests (Figs. 10.1 and 10.2) were conducted on photopolymers used for SLA 3D printing. These epoxy resins may contain antimony compounds up to 10% (by weight). Results of the study indicated that up to 3% of the total antimony contained in the material may be leached over the standard test duration of 20 h, generating a concern.

With the reduction of costs to a couple/a few hundred dollars, simply almost anyone can buy an SLA 3D printer. Even before the availability of these low-cost machines, the lead author was able to purchase SLA/DLP resins like MakerJuice [27] in the early 2010s when he was developing his own DLP machines. Thus, the control over these machines and their raw materials via regulations have become critical due to them being readily available to and accessible by large audiences. Besides the details mentioned above, sanding these SLA 3D printed parts may also release antimony in particular form to the air. This is another concern for worker safety and health, not so much for the environment, and is covered in the next section of this chapter. However, this concern also led to study of similar situations such as powder contamination caused by the early Binder-Jetting machines if proper care was not applied.

The SLA environmental work also drew additional collaborators from Centers for Disease Control's (CDC) National Institute of Occupational Safety and Health (NIOSH) Division (Aleks Stefaniak), causing a shift in the work towards the FDM/FFF processes.

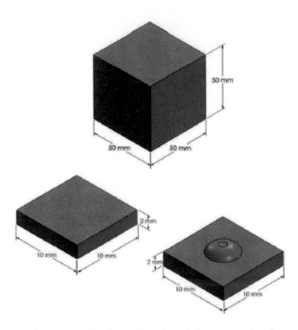

Fig. 10.1 SLA samples for the environmental study—the cross-sectional test (top) and leaching studies (bottom)—(left—cleaned and solvent-stripped, right—uncleaned and non-solvent stripped) [26]

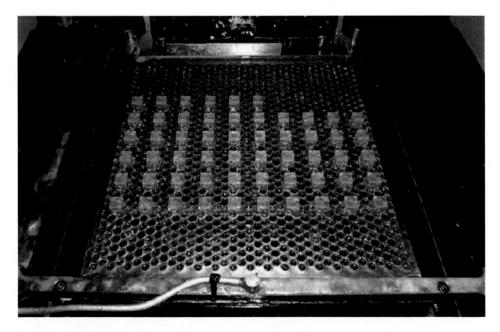

Fig. 10.2 SLA Viper machine printing leaching samples [26]

In the meantime, the lead author continued to work on sustainability and environmental issues with help from Paul Badger.

The recent environmental and sustainability work of the lead author centered on two different folds: conversion of used consumer plastics (shopping bags, water bottles) into 3D printing filament, and recycling of 3D printing waste. The former work started in the early 2010s, and the latter started in 2018. Following sections summarize both, with the most recent work not being available to discuss since it is still in-progress.

- The research team at RMU experimented with HDPE and LDPE materials which are from waste stock. A variety of consumer plastics including shopping and small bags were used in the effort. HDPE shopping bags constituted the most of the materials (Fig. 10.3). Shopping bags were shredded and then heated in silicone oil and cooled to give some rigidity (Fig. 10.4a), washed to remove the silicone oil, then chopped into small pieces (Fig. 10.4b) to be extruded in a Filabot™ machine (Fig. 10.5). Filaments made from shopping bags will have printability problems, due to the materials inability to stick on other surfaces. Thus, like the major chemical companies, the research team sought additives to improve the performance of the its HDPE filaments. A literature review indicated that the DOW Chemicals employed Fusabond® and Surlyn® for HDPE, both developed by DuPont. These additives are *Polyolefin (PE and PP)* compatibilizer resins. Adding Fusabond® as a compatibilizer improves mechanical properties to enable processing and direct recycling into the structure or into new applications [28]. Since these additives were not available in small quantities to purchase. The research team looked for other alternatives and found Taulman's T-Lyne material, an improved PE copolymer material with Surlyn®. Even this material became unavailable [29] after its discovery by the research team, the team continued to explore other compatibilizers and recently had some success in printing with HDPE-based printer filaments.

Fig. 10.3 Shopping bags

a   b

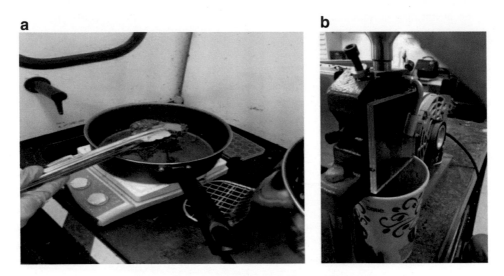

Fig. 10.4 **a** Heating HDPE in silicone oil. **b** Chipping HDPE chunks

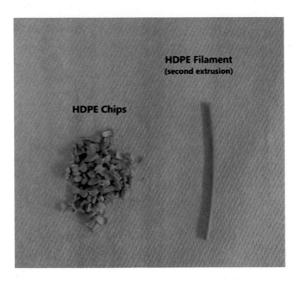

Fig. 10.5 A filament made by HDPE shopping bags

FFF 3-D printing is originally thought to be environmentally sustainable. However, significant amounts of waste can be generated employing this process. To improve its sustainability the lead author and his collaborators at CDC/NIOSH started a recycling project, acquiring multiple shredding and chipping equipment, Filabot™ filament makers, Airpath Coolers, and Spoolers also from Filabot™. The work centered around recycling

Fig. 10.6 PLA recycling workflow

the PLA waste generated in the RMU's Learning Factory. The following details the early progress as shown in Fig. 10.6:

- The first step in the process was chipping the 3D prints into small enough particles to be extruded. Upon the first extrusion process the team found out that the uneven particle size caused inconsistencies in the extrusion process that led to the filament to extrude in inconsistent diameters and would not be usable in a 3D printer. To fix this problem the team chipped the inconsistent filament again in order to obtain more uniform particle sizes to operate in a 3D printer. Once the second round of filament was being processed through the extruder, the team observed that it indeed was being extruded at a more consistent diameter that was usable. The final prints that were developed were acceptable but in no means perfect. To further its research with PLA the team acquired beads of a commercial material to see if perfectly consistent particle size would help, and this material was mixed with green recycled PLA to determine to improve the filament quality of the recycled material—Fig. 10.7.

Fig. 10.7 Efforts for
improving the quality of the
recycled PLA

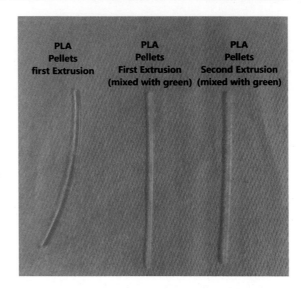

In Practice
SOLIDWORKS software offers a sustainability (Product *Life-Cycle Analysis (LCA)*)
analysis option under its "Evaluate" menu. After the "≫" expansion icon (upper
right of the software window) is selected, a set of analysis types will appear as
shown in the figure below including the "Sustainability" option. The following
example is based on molding a PET bottle with a diameter of approximately 2
inches and height of 3 inches. Following the figure is the analysis report generated
by the software. With this practice attempt, please analyze a part/product you have
designed using the same module and prepare a similar report.

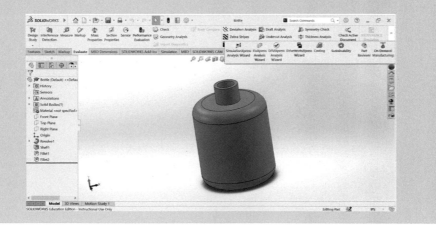

[company logo here]

[company name here]   [city, state here]   [company url here]

here

SOLIDWORKS
Sustainability Report

Model Name: Part1

Material: PET
Recycled content: 0.00 %
Weight: 0.11 lbs
Manufacturing process: Injection Molded
Surface Area: 43.16 in²
Built to last: 0.500 year
Duration of use: 1.0 year

■ Manufacturing Region
The choice of manufacturing region determines the energy sources and technologies used in the modeled material creation and manufacturing steps of the product's life cycle.

■ Use Region
The use region is used to determine the energy sources consumed during the product's use phase (if applicable) and the destination for the product at its end-of-life. Together with the manufacturing region, the use region is also used to estimate the environmental impacts associated with transporting the product from its manufacturing location to its use location.

Summary

Learn more about Life Cycle Assessment ◎

Sustainability Report

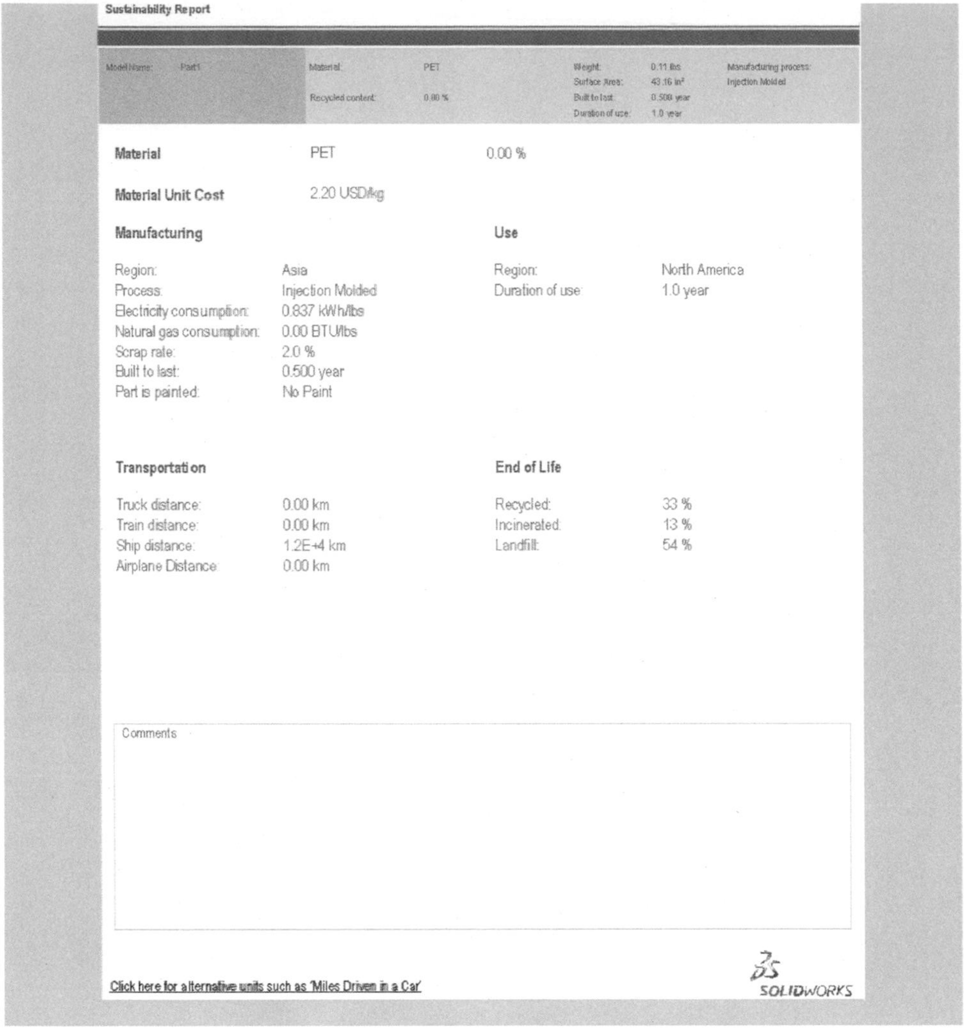
Sustainability Report

Model Name: Part1
Material: PET
Recycled content: 0.00 %
Weight: 0.11 lbs
Surface Area: 43.16 in²
Built to last: 0.500 year
Duration of use: 1.0 year
Manufacturing process: Injection Molded

Material	PET	0.00 %
Material Unit Cost	2.20 USD/kg	

Manufacturing

Region:	Asia
Process:	Injection Molded
Electricity consumption:	0.837 kWh/lbs
Natural gas consumption:	0.00 BTU/lbs
Scrap rate:	2.0 %
Built to last:	0.500 year
Part is painted:	No Paint

Use

Region:	North America
Duration of use:	1.0 year

Transportation

Truck distance:	0.00 km
Train distance:	0.00 km
Ship distance:	1.2E+4 km
Airplane Distance:	0.00 km

End of Life

Recycled:	33 %
Incinerated:	13 %
Landfill:	54 %

Comments

Click here for alternative units such as 'Miles Driven in a Car'

SOLIDWORKS

10.3   Health, and Safety Issues of 3D Printing and Additive Manufacturing

According to Hammer and Price [30], when the workers are assigned to machines to operate, they can do very little to change the adverse effects of those machines. However, they may be able to: recognize the effects associated with those machines, determine if the machines are in compliance with *Occupational Safety and Health Administration (OSHA)* or state standards, and if they are not, determine if any improvements can be made. They also defined a set of fundamental terms pertaining to the industrial safety and health area [30]:

- *Hazard*: Condition with a potential to cause injury (sudden and severe/acute) or sickness (gradually occurring over a time period/chronic).
- *Danger*: Expression of a relative exposure to a hazard.

- *Damage*: Severity of injury/sickness or the physical, functional, or monetary loss if an exposure to a hazard happens.
- *Safety*: Implication of freedom from hazards. It is practically impossible to entirely eliminate all hazards. It is a matter of relative protection through mitigation: via *engineering controls* (improving the source of the hazard, employing safety mechanisms, *personal protective equipment (PPE)*, or other technical means) or *management controls* (limiting exposure of the worker to the hazard in concern). The remedies mentioned above should reduce the level of the hazard to a point that it cannot do any harm.
  - *Descending order of preference* in dealing with hazards is given as: designs to eliminate/limit hazards, safeguard mechanisms, PPE, automatic warning devices, adequate procedures and personnel training.
- *Risk*: *Expression of possible loss over a specific period of time or during a number of* operational cycles. It may be indicated by *the probability of an accident times the damage in dollars, lives, or operating units.*

In the safety and health area, there are two main types of analysis: *hazard and risk analysis* (or assessment) and *accident/sickness analysis*, before and after the fact that some event with adverse effect has happened or is happening respectively. Following section covers the basics of hazard analysis including an example.

Each product, machine or process has certain inherent hazards, including a limited number of primary hazards, and a larger number of initiating and contributing hazards. These hazards can be identified by the two following approaches [30]:

- Theoretical possibilities can be added to a database from prior experiences.
- Theoretical aspects can be studied and then confirmed by actual experiences.

The following example explains the role of each hazard type. If we consider the catastrophic failure of a pressure vessel made of ordinary carbon steel, we will see that moisture played a starting role (*initiating hazard*) causing material to corrode and loose its strength (*contributing hazards*), leading to the rupture of the vessel (*primary hazards*). Primary hazard is the hazard type that can directly/immediately result in injury or death, release of materials, damage to the ambient including the machines and facilities, degradation of functional capabilities. In terms of the nature of the hazards, the following list covers most of them [30]:

- Biological Hazards including Bloodborne Pathogens
- Chemical Hazards
- Computers, Automation, and Robots
- Electrical Hazards
- Ergonomics Hazards
- Falling, Impact, Acceleration, and Lifting

- Fire Hazards
- Industrial Hygiene and Confined Spaces
- Mechanical Hazards and Machine Safeguarding
- Noise and Vibration Hazards
- Pressure Hazards
- Radiation Hazards
- Temperature Hazards
- Vision Hazards.

As mentioned earlier in the previous section, the lead author of this book started investigating the impact of nanomanufacturing and 3D printing in the mid-2000s on safety and health of the workers and engineers involved, producing a few scholarly works in the process [24–26]. The first publication was centered on a review of 3D printing (rapid prototyping) processes and associated safety and health issues, and presented at the 2010 RAPID/3D Scanning Conference. Table 10.4 illustrates a similar but up-to-date version of the assessment work [24]. This work was later expanded and published in Emerald's Rapid Prototyping Journal, in collaboration with an RMU chemist (Paul Badger) and an RMU environmental scientist (Daniel Short). This article followed a slightly approach, based on the three steps given below [25]:

- Review of relevant current OSHA regulations, *National Fire Protection Association* (NFPA) advisories, and hazards associated with rapid prototyping technologies.
- Risk assessment of rapid prototyping technologies under consideration.
- Review of potential chemical and environmental hazards by investigation of instrument manuals, Safety Data Sheets (*SDSs*), and toxicological literature searches.

During the Rapid Prototyping Journal work [25] the team identified the existence of antimony, a heavy metal used as a photo-initiator for the SLA process, changing the direction of the efforts to study leaching of that material [26]. A brief study on powder contamination of laboratory environments housing the Binder-Jetting technology was also conducted. Due to the collaboration request from CDC/NIOSH, the team started working on recycling, material degradation, and industrial hygiene issues of FFF printers since these printers are very common in educational environments including libraries, or homes. Two projects resulting from the RMU—CDC/NIOSH joint effort are summarized below [31, 32]:

- FFF 3-D printing is a valuable tool for engineering education. However, during the printing process, thermal degradation of the polymer filament releases small particles and chemicals, which are hazardous to human health including faculty, staff, and students populating the educational environment [31]. In this study, particle and chemical emissions from 10 different filaments made from virgin (never used) and recycled

Table 10.4 Hazard assessment for 3D printing processes

Hazard type	3D printing process details
Chemical hazards	• Common for liquid processing machines with easy access to the raw material container (i.e. DLP, SLA) • Chemicals used in post-processing or debinding are also a concern (i.e. BDM, FDM, SLA) • PPE's are required as well as SDSs
Confined spaces	• Inert gas mixtures used in sintering equipment (i.e. Binder-Jetting, BDM) may leak from their cylinders generating a confined space effect • Can be prevented by valve safety features
Electrical hazards	• All of the machines have electrical systems that are potentially hazardous. Issues are very rare due to well-set machine designs
Fire hazards	• Associated with liquid processing machines (i.e. DLP, SLA) as well as machines with powder-based raw materials (i.e. EBM, SLS/DMLS, SLM), also include explosion risks • Processing under vacuum and appropriate powder mitigation technologies exist (i.e. Binder-Jetting, EBM) • Raw materials need to be kept in flammable cabinets
Industrial hygiene	• Particle emissions and chemical releases in gas form are observed with solid processing machines (i.e. FDM/FFF) as well as powder releases from powder-based (i.e. Binder-Jetting/SLS), and chemical vapors from liquid processing machines (i.e. SLA, DLP) • Fume hoods and management controls can be used to limit exposure
Mechanical hazards	• Common for machines without enclosures (i.e. FFF) where users are exposed to the moving components • Most other machines should be protected by their enclosures and interlocks
Temperature hazards	• Common for machines without enclosures (i.e. FFF) where users are exposed to the hot ends/extruders/build plates • Most other machines should be protected by interlocks • Removal of recently made parts may require long cooling times (i.e. DMLS, SLS)
Vision hazards	• Includes laser and UV radiation (i.e. SLA/SLS/DMLS/SLM, DLP) • Machine enclosures and interlocks are effective in isolating users from these hazards

polymers were employed to print the same object at the *polymer manufacturer's recommended nozzle temperature (normal)* and *at a temperature higher than recommended (hot)* to simulate the real-world scenarios of a person intentionally or unknowingly printing on a machine with a changed setting. Emissions were evaluated in an RMU teaching laboratory using standard sampling and analytical methods. According to the mobility sizer measurements, particle number-based emission rates were 81 times higher, the proportion of *ultrafine particles (diameter <100 nm)* were 4% higher, and median particle sizes were a factor of 2 smaller for hot-temperature prints compared with normal-temperature prints (all with p-values <0.05). On the contrary, there was no difference in emission characteristics between recycled and virgin acrylonitrile butadiene styrene (ABS) and polylactic acid (PLA) filaments. Based on the test results, the team recommended reducing contaminant releases by using the following hierarchy of controls:

- Elimination/substitution by training students on principles of prevention-through-design, limiting the use of higher emitting polymers when possible.
- Engineering controls by using local exhaust ventilation via fume hoods to directly remove contaminants at the printer or isolating the printer from students.
- Management controls such as password protecting printer settings and establishing and enforcing adherence to a *standard operating procedure (SOP)* based on a proper risk assessment for the setup and use (limiting the use of temperatures higher than those specified for the filaments used).
- Good maintenance practices of printers.

• FFF 3-D printing is thought to be environmentally sustainable [32]. However, significant amounts of waste are generated from these 3D printers since they are widely available and affordable to a very large audience worldwide. Distributed recycling of plastics in homes, schools, and libraries to create feedstock filament may be the solution. Thus, the research team decided to study risks from exposures incurred during recycling and reuse of plastics used in these machines (Fig. 10.8). This study characterized contaminant releases from virgin (never used) and recycled plastics from filament production through the FFF process. Waste PLA and ABS materials were recycled to create filament. Virgin PLA, ABS, HDPE, LDPE, high impact polystyrene (PS), and polypropylene (PP) pellets were also extruded into filament. The release of particles and chemicals into educational settings was evaluated employing standard industrial hygiene methods. All tasks observed released particles that contained hazardous metals (like manganese) and with size capable of depositing in the gas exchange region of the lung, i.e., chipping of waste PLA and ABS (667–714 nm) and filament making (608–711 nm) and FFF 3D printing (616–731 nm) with waste and virgin plastics. All tasks observed released vapors, including respiratory irritants and potential carcinogens (benzene and formaldehyde), mucus membrane irritants (acetone, xylenes, ethylbenzene, and methyl methacrylate), and asthmagens (styrene, multiple carbonyl compounds). These data are imperative for incorporating risks of exposure to hazardous contaminants in EHS work as well as in future life-cycle evaluations to demonstrate the sustainability and circular economy potential of FFF 3-D printing in distributed spaces.

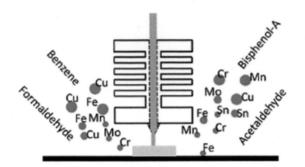

Fig. 10.8 FFF printer emissions [32]

10.4   Multiple Realities for IIoT and Industrial Training

Role of multiple realities has been growing in engineering applications and training including manufacturing. This section looks into extended reality (XR) via its components: augmented, virtual, and mixed or hybrid reality (AR, VR and MR/HR), their applications and associated hardware and software.

Researchers defined a spectrum indicating the level of involvement of the real and digital (virtual) worlds in a world of multiple realities, as presented in Fig. 10.9. At the edges of this spectrum are the real, and virtual worlds and anything in between is considered mixed (or hybrid) [33]. The virtual end of the spectrum is represented by virtual simulation (VS).

The following subsections of this chapter describe VS, AR, and VR technologies along with their applications including their role in training.

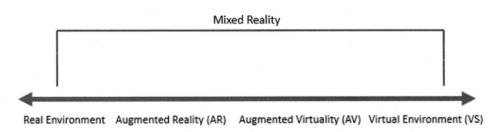

Fig. 10.9 Reality-virtuality spectrum [33]

10.4.1 Virtual Simulation

DDS's DELMIA [34] or Siemens' Product Life-Cycle Management (PLM) software [35] is a good example for generating a virtual environment—i.e. a digital factory with its machines also including *digital human models (manikins)*. This type of environment is called virtual simulation (VS). DELMIA software has recently been migrated into a cloud-based platform, 3DEXPERIENCE [36]. A comprehensive VS environment can be employed for design of products including complex ones, manufacturing processes and systems employed in making them, and their facilities. A CAD environment like CATIA originally ran on the background of DELMIA allowing users to design parts and complex products, and virtually test them with an associated Finite Element Analysis (FEA) application like SIMULIA. The very same environment facilitated inclusion of virtual machines, material handling mechanisms, storage facilities, tools, and humans for process and facility design for (i.e.) robotic spot welding of cars in a virtual assembly line, its material and people flow with a discrete-event simulation tool, along with safety and ergonomics analysis via manikins. Industrial robot companies such as FANUC [38] and ABB also have digital tools similar to DELMIA/ 3DEXPERIENCE but they are much less comprehensive. FANUC's simulation tools include modules such as HandlingPro, PalletPro, and PaintPro. Figure 10.10a presents a simulated machine tending operation in HandlingPro while Fig. 10.10b is illustrating a robotic palletizing operation in PalletPro. Both of these applications were employed in the lead author's "Robotics and Automation" course over a period of fifteen years. Robot programs produced in VS or off-line programming software are downloaded to the robots later for accomplishing work.

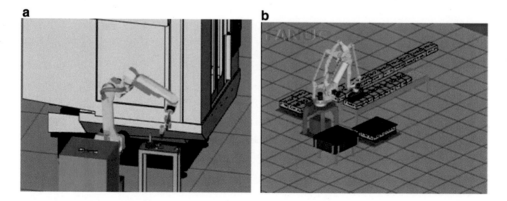

Fig. 10.10 **a** A pick and place operation of parts to be machined in a CNC [37], **b** palletizing of boxes filled with completed parts

10.4.2 Augmented Reality

AR is where the software application will superimpose or overlay alphanumeric, symbolic, or graphical information [39] onto the live view (of the real world) for enhancing the user's environment. AR can be accomplished in four ways as illustrated in Table 10.5.

AR devices and their technology originated in the early 1900s, where a targeting mechanism was patented to overlay a targeting reticle on a distant object, as it was focused on optical infinity [39]. This was followed by other sight reflector devices. The need for these devices stemmed from the human's inability to focus at two separate depth of fields. In the following decades, the systems of military fighter aircraft and helicopters became complex forcing pilots to process much more information such as critical flight information, sensor, and weapon systems data during the flight. This resulted in the first *modern heads-up (HUD) display* (in the 1950s), a transparent display mounted in front of the pilot preventing him/her to look down on the instrument panel. It also prevented the pilot to pay attention beyond the display outside the aircraft since it followed the same principle as the early *collimated devices focusing on the infinity* [39]. Further work in the 1960s led to the development of *helmet-mounted displays and sights* (HMDSs). While the HMDs presented similar information as HUDs, their sights also enabled the target to fall within the view of the device, allowing the pilot to steer missiles using his/her sight. Over the years additional devices have been developed to present *a single image to a single eye (monocular), single image to both eyes (biocular), separate viewpoint-corrected images to each eye (binocular).*

During the recent years, AR technologies have transitioned from defense and aerospace applications to gaming industry and industrial applications. AR devices existing today are categorized as:

- *See-through displays*—Fig. 10.11.
- *Video displays*—Fig. 10.12.

In addition, the AR hardware can be categorized based on their overall design and structure as:

- *Head Mounted Sights and Displays*: HUDs, HMDs
- *Holographic Displays*
- *Smart Glasses*
- *Handheld/Mobile Devices*: Smart Phones, Tablets, Laptops.

According to softwaretestinghelp.com, top 10 popular AR glasses are (ranked from 1 to 10): Oculus Quest 2, Lenovo Star Wars, MERGE AR/VR Headset, Microsoft HoloLens 2, Magic Leap One, Epson Moverio BT-300, Google Glass Enterprise Edition 2, Raptor AR headset, ThirdEye Generation, Kopin Solos [44]. Authors has had some experience with

Table 10.5 Definitions of different forms of AR

Definition	Example
• *With a target image (or marker)*: A static 3D image appears after the camera associated with the application recognizes a predetermined reference image (a marker)	An AR phone application utilizing RMU logo as a target image while a 3D scanned Robert Morris is used as a game asset 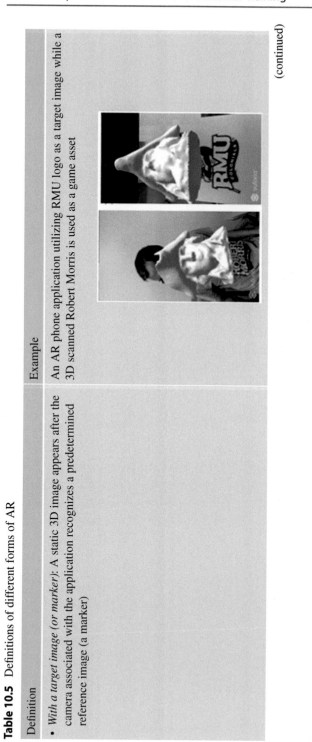

(continued)

Table 10.5 (continued)

Definition	Example
• *With a target image and animation(s):* An animated 3D image (or multiple images in a sequence) appears after the camera associated with the application recognizes a reference image. Users will not have control over the animation while operating the application	A computer AR application utilizing RMU logo as a target image leading to a jogging bear via a sequence of animated movements 

(continued)

Table 10.5 (continued)

Definition	Example
• *With a target image, animation(s), and control script.* One or more animated 3D images appear after the camera associated with the application recognizes a reference image. Users will have control over the animation during operation of the application based on the script via its "if" statements, calling and executing the animations	A computer AR app with a target image of a mechanical component for developing a 3D parts catalog. This example includes a virtual button (interactivity) as well as animation that rotates the part when pressed 

(continued)

Table 10.5 (continued)

Definition	Example
• *Without a target image (makerless)*: There are multiple different ways of making markerless AR including location-based markerless AR as in the well-known game of Pokémon GO. This method employs "the device's location and its sensor (camera) to unite the object with a point of interest" [40], without the prior knowledge of the real-world surroundings, including a predetermined marker. Static, animated, and interactive 3D images presented above are added to this type	A markerless spacecraft fighting AR game with interactivity (red laser trigger button) and multiple animations and special effects 

Fig. 10.11 See-through AR displays [41]—including a helmet-mounted display [42] and HoloLens 2 [43]

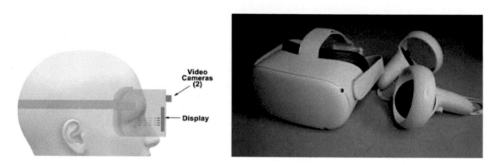

Fig. 10.12 Video AR displays [41]—where a camera-captured-live-view of the real-world is projected onto a display—Oculus Quest 2

Oculus Quest 2 and Hololens 2 via the lead authors "Augmented Reality Applications" course, and they are aware that Oculus Quest 2 has limited AR resources at the time this book was being written since it is the first version of Quest head-sets with AR ability unlike Hololens 2.

In terms of the AR development software, there are multiple paths developers can take including the popular and readily available game engines:

- *Unity Editor*: A popular method in the past was based on employing Unity Game Engine along with a Vuforia plugin. Unity now offers two new modules itself—*MARS* for AR development work and *AR Foundation* for publishing (deploying) the application in different environments including Android and iOS devices.
- *Unreal Game Engine for XR.*
- *Vuforia*: Vuforia *is* a now self-sufficient software tool, rather than being utilized as an AR plugin for the game engines.
- *ARCore*: "is a platform that enables Android application developers to quickly and easily build AR experiences into their apps and games. It can use your Android device's camera, processors, and motion sensors in order to serve up immersive interactions."

Google's ARCore allows AR apps to be developed in four different platforms (Android, Unity, Unreal, and iOS) before they are deployed in the Android environment [45].

- Apple also has multiple tools which can be used in AR development including *ARKit*, *RealityKit* as well as *Reality Composer* and *Reality Converter* [46].
  - ARKit handles the build process of AR apps.
  - RealityKit simulates and renders 3D image content for use in AR apps.
  - Reality Composer allows users to drag and drop virtual objects, assemble them directly into an AR scene, and animate them without requiring prior experience working with 3D assets [47]. Reality Composer has also a built-in AR library with ready-to-use AR assets.
  - Reality Converter app allows easy conversion, viewing, and customization of *USDZ 3D* objects on Apple hardware. USDZ is the file format of 3D and AR content on iOS devices without having to download special apps" [48].

10.4.3 Augmented Virtuality (Virtual Reality)

Augmented Virtuality or Virtual Reality (VR) is about immersion of the user into the virtual world. In a VR application, an engineer may be working on a virtual machine (as a part of the virtual world) to disassemble parts of it to study and improve its design, while interfacing with the environment via VR glasses. In a greater sense, an engineer can be a part of a digital factory fully developed in a VS environment, and almost interact with any parts of the factory's virtual operations—i.e. material handling and assembly, or disassembly (Fig. 10.13). This takes VS to another level as we arrive at VR.

Early VR headset development was also driven by military needs in a similar way to AR technology and is much newer due to the non-existence of computers. A 1982 U.S. Air Force project was focused on use of large VR headsets in simulating airborne system elements in a flight simulator. Again, more recent work was motivated by the entertainment and game sectors in order to make computing as ubiquitous as possible by eliminating the separation between the user and the computer. And most recently, we are seeing an acceleration of industrial and technical applications of VR. Similar to AR, VR also has applications in product, process, and systems design along with safety and hazard analysis, and training.

In terms of the development software, there are many possibilities including the popular and readily available game engines as well as some relatively unknown tools like Nintendo LABO kits:

- *Unity Editor.*
- *Unreal Game Engine for XR.*

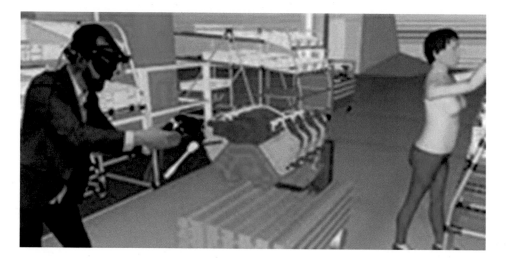

Fig. 10.13   A superimposed VR example [33]

- *Amazon Web Services (AWS) Sumerian Classroom*: Mainly a VR presentation software, this program has been recently discontinued, and the existing work can be moved to Babylon.Js.
- *Nintendo Switch Development Environment/LABO kits*: Toy Con Garage 04 modules of LABO kits allow VR game development and can be utilized with LABO VR glasses as well as the Nintendo Switch game console and its controllers [49]. These games can be used in teaching design and manufacturing concepts as well as being utilized for training of all sorts.

HMDs like Oculus Go, Quest and Quest 2 are common VR tools. According to pcg amer.com [50], the top VR head-sets are (ranked from 1 to 5): Oculus Quest 2, Valve Index, HTC Vive Pro 2, HP Reverb G2, and HTC Vive Cosmos Elite. These rankings reflect game performance of these tools, but can be translated to technical uses as well. Position and orientation information from these headsets can be interfaced to control other physical devices as well as user's activities in the VR environment. Additional hardware such as *Virtusphere, a 10-foot diameter device that rolls in place*—a device similar to a hamster's wheel, provides an infinite spherical plane where a user can walk in any direction. The enthusiasts of VR—can have access to it using cardboard holders of smart-phones and associated VR apps, in a similar way to Nintendo Switch game consoles placed in cardboard LABO VR glasses as shown in Fig. 10.14. Figure 10.15 presents a VR game developed by the lead author's students at Robert Morris University employing LABO Toy Con Garage. Other immersive environments also include 360° projection of information in room walls and CAVE (computer-assisted virtual environments).

Fig. 10.14 Nintendo LABO VR glasses embedded with the Switch game console and its controllers [49]

Fig. 10.15 A Nintendo VR game [49]

10.4.4 Comparing and Contrasting XR Hardware

A large number of characteristics can be considered including those of audio and video displays, while reviewing and choosing XR hardware, no matter if it is for AR or VR. Naturally the *costs* and technical specs like *RAM or storage capability* are some of the important factors. Headsets can vary from a few hundred dollars (Oculus Quest 2) to a few thousand dollars (HoloLens 2). These devices can also have varying RAM and storage capacities:

- Oculus Quest 2–6 GB RAM with a storage capacity of 128 GB or 256 GB available.
- Hololens 2–4 GB RAM with a storage capacity of 64 GB.

Even though it is impossible to compare all of these devices and their hardware within the scope of this chapter, the authors decided to list the different factors involved in only video display technologies and their performance—Table 10.6 [51]. A general comparison based on the ocularity of the optics are also given below in Table 10.7 in addition to the technologies found in XR hardware, for possible future study for the interested readers [52].

10.4.5 Industrial and Educational Applications

Applications of XR tools are constantly evolving within the engineering field and beyond. One company that has embraced XR to assist in product simulation and design analysis is the Ford Motor Company [53]. Virtual reality is utilized to explore the vehicle design on both the exterior and interior to look over the structure and how various systems interact with one another. The aim of this effort is to study the design feasibility along with the employee's safety while working on the production line [53]. With the application of VR technologies, the injury rates on the production line have decreased by 70% [53]. This technology is also being used to train people in service for Mustangs without having a physical model of the electric Mustang Mach-E in the factory [54].

The use of XR expands beyond the automotive industry and goes into aerospace manufacturing with the Lockheed Martin Corporation [53]. This corporation is responsible for NASA's Orion Multi-Purpose Crew Vehicle. This spacecraft was originally constructed in virtual space, which allowed the project team to review the overall design and system for any possible issues before constructing the prototype [53]. VR is more readily utilized in the product design phase because it is the most important task within the manufacturing realm [55]. This process is also referred to as *virtual manufacturing (VM)*. According to George Leopold, AR has also allowed this company to increase its productivity by more than 40% while saving millions [56].

One of the first areas that utilized AR is architecture education [57]. This is because architecture walkthroughs were the baseline application. The architecture industry also uses AR for spatial analysis, design review, and problem identification [57]. It is also employed to add objects virtually into a physical space to visualize how an object works before buying an object. Another application of AR in education can be based on reviewing a 3D model of an object before making a physical prototype. This reduces the material wasted if the object has some problems with its design, by employing a technology such as *the Looking Glass*—a holographic display [58]. The next section of this paper focuses on this hardware and its relevant uses in design and manufacturing.

One area that has been interfacing with engineering in some aspects is the medical field [58]. VR has become common within the curriculum for medical students to optimize their learning, and is extended to biomedical engineering including its device development area. This type of learning has proven to be a better experience for the students and provides

Table 10.6 Technologies available for imaging and displaying along with associated factors and characteristics [51]

Imaging and display technology	Light performance factor	Display properties and characteristic	Purposes of optics	Optical architecture concept	Waveguide type
Liquid crystal displays (LCD)	Lumen	Spatial resolution	Collimation of light	FOV	Holographic
Organic light-emitting diode (OLED)	Luminous flux	Pixel pitch	Magnification of the image	Interpupillary distance (IPD)	Diffractive
Active matrix organic light-emitting diodes (AMOLED)	Luminous Intensity	Fill factor	Relay of light patterns	Eye relief	Polarized
	Candela	Persistence		Eye pupil	Reflective
	Luminance	Latency		Eye box	
Digital light projector (DLP)	Illuminance	Response time			
	Brightness	Color Gamut			
Liquid crystal on silicon (LCoS)		Contrast			

Table 10.7 Modes of ocularity and their advantages and disadvantages [52]

Ocularity mode	Advantages	Disadvantages
Monocular display	"Low weight, small form factor, least distracting, easiest integration, least computational overhead, easiest alignment"	"Possibility of binocular rivalry and eye-dominance issues, small field of view (FOV), no stereo depth cues, asymmetric mass loading, reduced perception of low-contrast objects, no immersion"
Biocular	"Less weight than binocular, no visual rivalry, useful for close proximity training tasks requiring immersion, symmetric mass loading"	"Increased weight, limited FOV and peripheral cues, no stereo depth cues, often lens is larger to accommodate larger eye box"
Binocular	"Stereo images, binocular overlap, large FOV, most depth cues, sense of immersion"	"Heaviest, most complex, most expensive, sensitive to alignment, computationally intensive operation"

a greater competence of the skills learned rather than through standard lecture and text-based education. This was seen through a study conducted on medical students across 8 medical schools in Pakistan [59].

XR has also been emerging within the trades, such as welding or spray painting [57]. VRTEX 360 Weld Simulator is a VR-based curriculum that can be used to make education more efficient and less expensive. This training style allows students to have an experience that is consistent with industry criteria for methodology and evaluation [57].

10.4.6 Applications and Software Tools of the Looking Glass Hologram

There are other tools in the MR space that include the Looking Glass hologram, 360° projectors, and smartphones [58]. The Looking Glass is a hologram projector that portrays images in 3D. It has the capabilities to be integrated with smartphones and computers through a variety of apps and software. To use the Looking Glass with a computer, the computer must have *Holoplay Service* installed and the display settings set as an extended screen. Once the proper software is downloaded, the user can test the hologram by navigating to the Looking Glass Factory to the Test Holoplay Service Connection feature. The image that appears on the hologram should look like the one in Fig. 10.16. The model holograms on the company's website serve as an example of the potential uses for the device. One specific model allowed the user to interact with the hologram with a hand tracker (a motion sensor) for moving the objects around as shown in Fig. 10.17.

The Looking Glass is compatible with specific software tools including *Holoplay Studio*, a 3D model importer, and plug-ins for various programs [58]. The model importer allows the user to import specific 3D XR files that are in *.GLTF, .GLB, and .OBJ* formats,

Fig. 10.16   Reference hologram image for the looking glass [58]

Fig. 10.17   A CAD image is being manipulated by a motion sensor (leap motion) tracking user's hand [58]

and interact with the model. The users will be able to move the image and zoom in to obtain a different perspective on the part as seen in Fig. 10.17. This program allows the user to view their model in 3D space without the need to 3D print it and waste material, thus essentially saving both time and money. There are also plug-ins for various applications including Unity Editor, Blender, and the Unreal Game Engine for developers. These plug-ins allow the user to view their design and its virtual space, and observe them from a 3D perspective. It can also be used to convert .STL files into .GLTF files to be displayed. Likewise, the Unity plug-in or the Unreal Engine plug-in can be used in technical applications as well as game design and development.

For the user to interact with their desired holograms on the Looking Glass, they can utilize a hand motion tracker known as the *Leap Motion* controller. It has multiple applications including gaming purposes [60]. However, it has a great potential to spread far beyond just the gaming experience. The Leap Motion utilizes an "*infrared light-based stereoscopic camera* [60]" to track the user's hand movements with an algorithm as seen in Fig. 10.18. As MR has been utilized in different training programs such as welding, the Leap Motion has the potential to be implemented in these scenarios where delicate hand movements are mandatory.

These programs and technology easily convert into tools to be used for educational purposes in teaching aspiring engineers [58]. Students will be able to design products and then view them in an MR space. They will be able to interact with their designs with the Leap Motion and view them at the same time with the Looking Glass (Fig. 10.17). Not only does the Looking Glass have educational potential, but it also has potential in the design and marketing business. It will allow innovators to design their product and present their prototype with it. With the use of a virtual prototype, businesses will save time, money, and resources in the design process which then can be allocated elsewhere. However, a disadvantage of the Looking Glass hologram project is that it is limited in size, and works best when viewed at a specific angle. Another technology that can be applied in the MR realm is the application of the 360° projector to enhance the workspace. This projector gives the user a fully immersive experience without the need for a headset, they will be able to "provide the helper with a view of the task space independent of the workers orientation" [61].

Furthermore, tablets and smartphones have the potential to implement AR with different technologies for multiple purposes. In their article, Figueiredo and co-authors introduced a prototype known as the *EducHolo*, which enabled the visualization of holograms with tablets [62]. Their purpose was to provide a different perspective of 3D models of parts for mechanical engineering students [62].

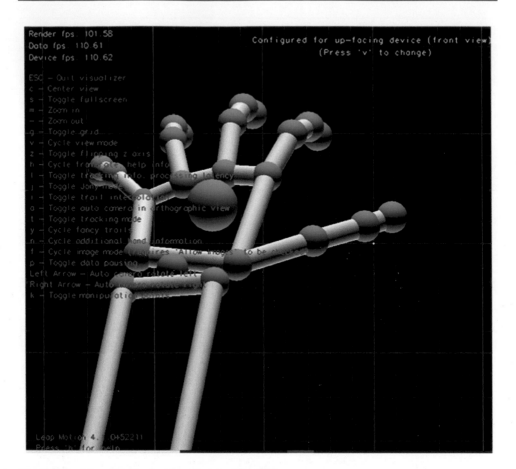

Fig. 10.18 Algorithm for hand motion tracking [58]

10.4.7 The Future of Multiple Realities

The implementation of XR is becoming an element of engineering education to both help students in the classroom and prepare them for their future careers, as this technology is also becoming more readily used in a multitude of industries. Its implementation in the automotive industry will conserve materials and aid in the design and development of vehicles, all the whilst achieving a safer workplace for its employees. The approach by Ford can be adapted to provide an inclusive experience for aspiring engineers. XR has a promising role in engineering education, as it offers students the opportunity to experience the manufacturing process. They would be able to design a product and test it in simulated environments, providing them with the necessary experience in a cost-efficient manner.

Society is just starting to utilize AR, and the uses will continue to grow over the coming years. VR has spread from the gaming industry, and more systems will dive deeper into this use of headsets and hand tracking to optimize the consumer's experience. Another direction AR might be going in is separating the optics from the display. DARPA has developed this type of system in which the "magnifying optics are completely removed from the eyewear and are instead integrated into an advanced contact lens" [63]. These contact lenses can be used within the system and as corrective lenses when the system is not in use. Another aspect that will be enhancing the VR experience is the use of 3D sound. This is the use of some spatial sound solutions to create a more immersive experience [63]. The use of holograms is expanding the possibilities of collaboration amongst varying groups, as multiple groups can be working on a project together through the hologram [64]. This allows multiple users to be working in AR environments without the use of headsets.

Review Questions

1. What are the three main cost components of a 3D printing process?
2. What factors do contribute to pre-processing, build, and post-processing costs of a 3D printing process?
3. What are the five secret costs of post-processing?
4. If you are to summarize cost modeling and estimation efforts, how can you generalize works from the literature that was included in this chapter?
5. What are the five steps proposed by GE Additive to make an AM business case? Explain each briefly.
6. Prepare a mind map covering sustainability issues in 3D printing and additive manufacturing.
7. Prepare a mind map covering environmental issues in 3D printing and additive manufacturing.
8. List and define the fundamental terms used in the health and safety area.
9. What are two types of analysis employed in the health and safety area? Define each concisely.
10. What are two approaches used in identifying hazards?
11. Define the terms, "initiating hazard, contributing hazard, and primary hazard".
12. Prepare a mind map associating 3D printing processes and their hazards (Hint: Expand the information given in Table 10.4 if necessary).
13. Which 3D printing industrial hygiene issues were covered in the chapter? Explain each one briefly.
14. What type of engineering and management controls can be used in dealing with 3D printing and additive manufacturing safety and health issues?
15. Define the spectrum of multiple realities with a sketch.
16. Define virtual simulation (VS).
17. Define augmented reality (AR).
18. Define virtual reality (VR).

19. What is the difference between the terms, extended reality (XR), mixed reality (MR), and hybrid reality (HR)?
20. Prepare a mind map covering the uses of virtual simulation in an organized fashion.
21. What are different uses of AR in industry?
22. List the four different types AR covered in the chapter and explain each one with an example.
23. What are different uses of VR in industry?
24. Prepare a mind map covering the AR hardware and software.
25. Prepare a mind map covering the VR hardware and software.
26. Define monocular, biocular, and binocular displays. Compare and contrast each one through its advantages and disadvantages (Hint: Review Table 10.7).
27. What is virtual manufacturing? Define and give and give an example for it (Hint: It can be found in the chapter).
28. List and explain the alternative display tools available for engineering design and manufacturing activities with one example each.

Research Questions

1. Update the cost figures given in the first In Practice section and complete the cost study with the new figures.
2. Update the impact of the inflation figures on the FFF process using current statistical data (Hint: Locate the information that follows the first In Practice section).
3. Conduct a literature review to find an AM simplified supply chain example similar to the one included in the chapter, and summarize it using Table 10.3 as a reference.
4. Please modify the second In Practice section's requirement to include 3D printing as the manufacturing method (Hint: You will need to learn how define 3D printing in the SOLIDWORKS Sustainability analysis feature).
5. Find a case study that pertains to safety and health issues in digital manufacturing. Summarize the case in two paragraphs, also identifying its initiating, contributing, and primary hazards.
6. If you have access to a Nintendo Switch console, LABO Kits, and the Toy Con Garage 04 module (disk), prepare a VR learning game focusing on digital manufacturing. 3D print the game interface to be housing the Nintendo Switch console and its game controller.
7. Review the contents of Table 10.6 and expand it for preparing a PowerPoint presentation targeting technologies available for imaging and displays.

Discussion Questions

1. Explore the features of the ANSYS Granta Edupack software's Eco Audit tool and compare it to the SOLIDWORKS Sustainability feature.

2. Discuss the impact of alternative displays such as holograms or 360° projection on engineering design and manufacturing efforts. Limit your work to two pages and use appropriate referencing methods.
3. Discuss the future of collaborative work in engineering and in engineering education projects, while focusing on the tools covered in this chapter. Limit your work to two pages and use appropriate referencing methods.
4. Discuss the future of multiple realities in digital manufacturing in a single page brief. Use appropriate referencing methods.
5. Discuss the future of digital manufacturing in a three-page article. Use appropriate referencing methods.

References

1. Sirinterlikci, A., Erdem. E. (2019). Unpublished works–A Cost Estimation Study of Fused Filament Fabrication (FFF) Parts.
2. Ajay A., Rathore, A.S., Song C., Zhou C., and Xu W. (2016). Don't forget your electricity bills! An empirical study of characterizing energy consumption of 3D Printers, Proceedings of 7th ACM SIGOPS Asia-Pacific Workshop APSys, ACM, Hong Kong, China, pp. 7–14.
3. Piili H., Happonen A., Vaisto T., Venkataraman V., Partinen J., and Salminen A. (2016). Cost Estimation of Laser Additive Manufacturing of Stainless Steel, In Proceedings of 15th Laser Materials Processing Conference, Elsevier B.M., Lappeenranta, Finland, Amsterdam, The Netherlands, pp. 388–396.
4. Hitch J., Five Secret Costs of 3D Printing—New Equipment Digest [online], located at at: https://www.newequipment.com/research-and-development/five-secret-costs-3d-post-proces sing, accessed February 5, 2018).
5. Dong Y., Bao C., and Kim, W.S. (2018). Sustainable Additive Manufacturing of Printed Circuit Boards', Joule, 2 (4), pp. 579–582.
6. Winters, B.J. and Shepler D. (2018). 3D printable optomechanical cage system with enclosure. HardwareX, 3, pp. 62–81.
7. Zhang, C., Anzalone, N.C., Faria, R.P., and Pearce, J.M. (2013). Open-source 3D-printable optics equipment. PloS one, 8 (3), e59840.
8. Mostafa, K.G., Montemagno, C., and Qureshi, A.J. (2018). 'Strength to cost ratio analysis of FDM Nylon 12 3D Printed Parts, Proceedings of 46th SME North American Manufacturing Research Conference, NAMRC 46, Texas, USA, pp. 753–762.
9. Franchetti, M and Kress, C. (2017). An economic analysis comparing the cost feasibility of replacing injection molding processes with emerging additive manufacturing techniques, The International Journal of Advanced Manufacturing Technology, 88 (9–12), pp. 2573–2579.
10. Baumers, M, Dickens P, Tuck C and Hague R. (2016). The cost of additive manufacturing: machine productivity, economies of scale and technology-push, Technological forecasting and social change, 102, pp. 193–201.
11. Sirinterlikci, A., Uslu, O., Behanna, N., Tiryakioglu, M. (2010). Preserving Historical Artifacts through Digitization and Indirect Rapid Tooling, International Journal of Modern Engineering, 10 (2), pp. 42–48.

12. Zhang, Z. and Wang, S. (2017). Patient Decision Making for Traditional vs. 3D Printing-Based Meniscus Transplantation, Proceedings of Industrial and Systems Engineering International Conference, Pittsburgh, PA.
13. Fernandez-Vicente, M, Escario Chust, A., and Conejero, A. (2017). Low cost digital fabrication approach for thumb orthoses, Rapid Prototyping Journal, Vol: 23 (6), pp. 1020–1031.
14. Smektala, T. (2016) Low cost silicone renal replicas for surgical training–technical note, Archives Espania Urologica, 69 (7), pp. 434–4366, located at: https://pubmed.ncbi.nlm.nih.gov/27617553/, accessed September 20, 2022.
15. Cameron, T. and Gordon, A. (2017). In-plane cost analysis considering material loading limitations, Proceedings of AIAA Space and Astronautics Forum and Exposition, SPACE 2017, Orlando, FL.
16. Feldmann, C. and Pumpe A. (2017). A holistic decision framework for 3D printing investments in global supply chains, Transportation Research Procedia, 25, 2017, pp. 677–694.
17. Lan, H. and Ding, Y. (2007). Price Quotation Methodology for Stereolithography Parts based on STL model, Computers & Industrial Engineering, 7 (52), pp. 241–256.
18. Sirinterlikci, A. (2015), Unpublished RAMP2 Program Report—Employment of AM in Fluid Conditioning Product Development.
19. King J., The True Cost of Running a Desktop 3D Printer 3D Print Headquarters, located at: http://3dprinthq.com/cost-running-desktop-3d-printer/, accessed August 21, 2018.
20. Electric Rates. Pittsburgh Electricity Prices, located at: http://www.electricrate.com/2012/06/pittsburgh-electricity-prices/, accessed July 5, 2018.
21. Bureau of Labor Statistics Employment News Release, Employment Cost Index., located at: http://www.bls.gov/news.release/eci.toc.htm/, accessed December 5, 2017.
22. US Inflation Calculator. Consumer Price Index, located at: http://www.usinflationcalculator.com/inflation/consumer-price-index-and-annual-percent-changes-from-1913-to-2016/, accessed March 7, 2018.
23. ROI Playbook |GE Additive, located at: https://www.ge.com/additive/roi-playbook, accessed September 2, 2022.
24. Sirinterlikci, A., Artieri, B., Story, C., Coughlin, C., Jonnet, M., Eaton, B. (2010), Environmental, Health, and Safety Issues in Rapid Prototyping, SME Rapid/3D Imaging Conference, Anaheim, CA.
25. Sirinterlikci, A., Short, D., Badger, P.D., Artieri, B. (2015), Environmental, Safety, and Health Issues in Rapid Prototyping", Rapid Prototyping Journal, 21 (1), pp. 105–110, https://doi.org/10.1108/RPJ-11-2012-0111.
26. Sirinterlikci, A., Short, D., Volk, D., Badger, P.D., Melzer, J., Salerno, P., 3D Printing (Rapid Prototyping) Photopolymers: An Emerging Source of Antimony to the Environment, 3D Printing and Additive Manufacturing Journal, 1(1), https://doi.org/10.1089/3dp.2013.001.
27. MakerJuice Standard Red Resin, located at: https://www.matterhackers.com/store/3-d-printer-resins/makerjuice-g+red-resin-1-liter, accessed September 20, 2022.
28. Polymer Compatibilizer Improves the Value of Blended or Recycled Plastics, located at: https://www.dow.com/en-us/insights-and-innovation/product-news/polymer-compatibilizer-for-recycling, accessed September 20, 2022.
29. T-Lyne Filament, located at: https://www.matterhackers.com/store/l/taulman-t-lyne-flexible-filament-285-1Ib/sk/MFKRF6LU, accessed September 20, 2022.
30. Hammer, W., Price, D. (2000) Occupational Safety Management and Engineering 5th Edition, Hoboken, NJ: Prentice Hall International Series in Industrial and Systems Engineering.
31. Stefaniak, A., Bowers, L., Cottrell, G., Erdem, E., Knepp, A., Martin, S., Pretty, J., M. Duling, M., Arnold, E., Wilson, Z., Krider, B., LeBouf, R.F., Virji, A., Sirinterlikci, A. (2021) Use of 3-dimensional printers in educational settings: The need for awareness of the effects

of printer temperature and filament type on contaminant releases", ACS Chemical Health and Safety Journal, 28(6), pp. 444–456, https://doi.org/10.1021/acs.chas.1c00041.

32. Stefaniak, A., Bowers, L., Cottrell, G., Erdem, E., Knepp, A., Martin, S., Pretty, J., M. Duling, M., Arnold, E., Wilson, Z., Krider, B., LeBouf, Fortner, A., R.F., Virji, A., Sirinterlikci, A. (2022), Towards Sustainable Additive Manufacturing: The Need for Awareness of Particle and Vapor Releases During Polymer Recycling, Making Filament, and Fused Filament Fabrication 3D Printing, Resources, Conservation, & Recycling, 176, 105911, https://doi.org/10.1016/j.res conrec.2021.105911.

33. Milgram, P. and Fumio, K. (1994). A taxonomy of mixed reality visual displays, IEICE Transactions on Information and Systems, 77 (12), pp. 1321–1329.

34. DELMIA Products, located at: https://www.3ds.com/products-services/delmia/products/, accessed February 18, 2022.

35. Siemens Digital Industries Software, located at: https://www.plm.automation.siemens.com/glo bal/en/, accessed February 18, 2022.

36. 3DEXPERIENCE Cloud Platform, located at: https://www.3ds.com/3dexperience/cloud? utm_medium=cpc&utm_source=google&utm_campaign=202201_glo_sea_en_op51508_labl_ brand_nam_phrase&utm_term=3dexperience-phrase&utm_content=search&gclid=EAIaIQ obChMIn_2QrPOZ9gIVRJyzCh0jFg90EAAYASAAEgJYQPD_BwE, accessed February 18, 2022.

37. Sirinterlikci, A. and Al-Jaroodi, J. (2018), Industrial applications of AR, Digital Bridge Conference, Pittsburgh, PA.

38. Robot Simulation Software—FANUC ROBOGUIDE, located at: https://www.fanucamerica. com/products/robots/robot-simulation-software-FANUC-ROBOGUIDE, accessed February 18, 2022.

39. Aukstakalnis, S., *Practical Augmented Reality 1st edition*. Crawfordsville, IN: Addison-Wesley pp. 2.

40. Markerless AR, located at: https://www.marxentlabs.com/what-is-markerless-augmented-rea lity-dead-reckoning/, accessed February 18, 2022.

41. Aukstakalnis, S., Practical Augmented Reality 1st edition. Crawfordsville, IN: Addison-Wesley pp. 6.

42. Aukstakalnis, S., Practical Augmented Reality 1st edition. Crawfordsville, IN: Addison-Wesley pp. 290.

43. HoloLens2, located at: https://www.amazon.com/, accessed March 31, 2022.

44. Best Augmented Reality Glasses, located at: https://www.softwaretestinghelp.com/best-augmen ted-reality-glasses/, accessed Feb. 18, 2022.

45. ARCore, located at: https://developers.google.com/ar/, accessed February 18, 2022.

46. ARKit, located at: https://developer.apple.com/augmented-reality/, accessed February 18, 2022.

47. Reality Composer, located at: https://docubase.mit.edu/tools/reality-composer/, accessed February 18, 2022.

48. Ultimate Guide to USDZ File Formats–Marxent Labs, located at: https://www.marxentlabs. com/usdz-files/, accessed February 18, 2022.

49. Sirinterlikci, A., Mativo, J., Pham, J. (2020), Using Nintendo Switch Development Environment to Teach Game Development and Virtual Reality, ASEE Annual (American Society for Engineering Education) Conference and Exposition, ONLINE.

50. Best VR Headsets, located at: https://pcgamer.com/best-vr-headsets/, accessed March 31, 2022.

51. Aukstakalnis, S., Practical Augmented Reality 1st edition. Crawfordsville, IN: Addison-Wesley pp. 58–75.

52. Aukstakalnis, S., Practical Augmented Reality 1st edition. Crawfordsville, IN: Addison-Wesley pp. 57.

53. Aukstakalnis, S., Practical Augmented Reality 1st edition. Crawfordsville, IN: Addison-Wesley pp. 253–261.
54. M. Dearborn, How Ford, Boscho are Using Virtual Reality to Train Technicians on All-Electric Mustang Mach-E, Ford Media Center, located at: https://media.ford.com/content/fordmedia/fna/us/en/news/2020/02/14/ford-bosch-virtual-reality-technician-training-mustang-mach-e.html/, accessed February 18, 2022.
55. Choi S., Jung K., and Noh, D. (2015). Virtual Reality Applications in Manufacturing Industries: Past Research, Present Findings, and Future Directions, Sage Journals, 23 (1), located at: https://doi.org/10.1177/1063293X14568814/, accessed February 22, 2022.
56. Leopold, G. (2019), Lockheed Martin Embrace AR on the Shop Floor, Scope AR, located at: https://www.scopear.com/lockheed-martin-embraces-ar-on-the-shop-floor/, accessed February 18, 2022.
57. Aukstakalnis, S., Practical Augmented Reality 1st edition. Crawfordsville, IN: Addison-Wesley pp. 300–309.
58. Sirinterlikci, A., Wolfe, A., Farroux, L., (2022), Use of Interactive Digital Tools in Product Design and Manufacturing, American Society for Engineering Education (ASEE) Annual Conference and Exhibition, Minneapolis, MN.
59. Sattar, M., Palaniappan, S., Lokman, A., Shan, N., Khalid, U., and Hasan, R. (2020). Motivating Medical Students Using Virtual Reality Based Education, International Journal of Emerging Technologies in Learning (iJET), 15 (2), Jan. 29, 2020, located at: https://www.learntechlib.org/p/217172/, accessed February 18, 2022.
60. Wozniak, P., Vauderwange, O., Mandal, A., Javahiraly, N., and D. Curticapean, D. (2016), Possible applications of the LEAP motion controller for more interactive simulated experiments in augmented or virtual reality," in Proceedings of SPIE 9946, located at: https://doi.org/10.1117/12.2237673/, accessed February 21, 2022.
61. Rasmussen, T.A. and Gronbak, K. (2019). Tailorable Remote Assistance with Remote Assist Kit: A Study of and Design Response to Remote Assistance in the Manufacturing Industry, Researchgate, located at: https://www.researchgate.net/profile/TroelsRasmussen/publication/335436879_Tailorable_Remote_Assistance_with_RemoteAssistKit_A_Study_of_and_Design_Response_to_Remote_Assistance_in_the_Manufacturing_Industry/links/5d712bc2299bf1cb808912bf/Tailorable-Remote-Assistance-with-RemoteAssistKit-A-Study-of-and-Design-Response-to-Remote-Assistance-in-the-Manufacturing-Industry.pdf/, accessed February 18, 2022.
62. Figueiredo, M.J.G., Cardoso, P.J.S., Gonçalves, C.D.F., and Rodrigues, J.M.F. (2014), Augmented Reality and Holograms for the Visualization of Mechanical Engineering Parts, 18th International Conference on Information Visualization, pp. 368–373, located at: https://doi.org/10.1109/IV.2014.17, accessed February 22, 2022.
63. Aukstakalnis, S., Practical Augmented Reality 1st edition. Crawfordsville, IN: Addison-Wesley pp. 356–360.
64. Mavrikios, D., Alexopoulos, K., Georgoulias, K., Makris, S., Michalos, G. Chryssolouris, G. (2019), Using Holograms for visualizing and interacting with educational content in a Teaching Factory, Procedia Manufacturing, 31, located at: https://doi.org/10.1016/j.promfg.2019.03.063/, accessed Feb. 22, 2022.

Practicum

11

Digital manufacturing practice can be simulated in an educational setting by employing term-long projects. The lead author frequently utilizes two half-term long team projects in his "Rapid Prototyping and Reverse Engineering" course. Goal of the first one is to mimic product development practice, following a series of activities starting from conception to prototyping. A variety of objectives including modular construction toy development, developing tools for handicapped persons for helping improve their quality of life as well as employability are realized in the first term-long project. The second term-long project can be about reverse engineering an FFF 3D printer or a common mechatronic wind-up toy. As the student team assembles an FFF 3D printer, the members also follow the principles of reverse engineering presented in Chap. 9 of this book, preparing a report for it as if they reverse engineered the 3D printer. Following section sketches requirements of example product development and reverse engineering projects.

11.1 Product Development and Manufacturing Project

Student teams made from 4 to 5 students are tasked to design a construction toy (modular toys similar to LEGO™, with at least 6 distinct modules). The teams have to follow the steps given below during the development process and record the results for their reporting in a MS PowerPoint presentation:

- Developing a problem statement of their toy concept including its realistic constraints such as age and cultural appropriateness.
- Brainstorming and producing alternative design solutions.
- Analyzing the alternative designs based on criterion the team develops and choose the best alternative.

© The Author(s), under exclusive license to Springer Nature Switzerland AG 2023
A. Sirinterlikci and Y. Ertekin, *A Comprehensive Approach to Digital Manufacturing*,
Synthesis Lectures on Mechanical Engineering,
https://doi.org/10.1007/978-3-031-25354-6_11

- Developing the system level design based on the best alternative selected.
- Detail designing each toy piece with its geometry, dimensions/tolerances, material and manufacturing process selection etc.
- Cost estimation study.
- Conducting at least one type of engineering analysis or testing (FEA analysis mimicking a drop or a bite test, SolidWorks sustainability analysis for the product life-cycle, form-fit-function checking).
- Evaluating the analysis or test results and adding safety standards information relevant to the product type or its specifications (updating its constraints).
- Building a physical prototype through 3D printing or a virtual prototype through animation (often done by a SolidWorks motion study).
- Preparation of user instructions to be included with the toy.

Figure 11.1 illustrates a part of a concept developed by a student team to incorporate motor movement into an existing commercial toy, ZOOB. This idea led to a product line of toys under the name of ZOOB Motor Company [1] marketed by the company, Infinitoy. The student team modified existing ZOOB components to facilitate electrical motor actuation. The original ZOOB toy components (Fig. 11.2) were modeled based on DNA nucleotides and was allowing only manual movement.

Another construction toy project example is given in Fig. 11.3 and includes different assemblies made by the concept toy [1]. Often students modified existing toys or are inspired by them. Intellectual property laws are covered earlier in the course and revisited during the course of these product development projects.

Finally, Fig. 11.4 illustrates a system level design chart prepared by a student design team for their project, including two different categories of parts; rods and connectors. While Fig. 11.5 presents rod category parts which are included under the subassembly 1 in the system level design, connectors are given in Fig. 11.6 along with an assembly of multiple rotating connectors (subassembly 2). Figures 11.5 and 11.6 also hint that 3D printing presents an excellent medium for construction toy development, along with

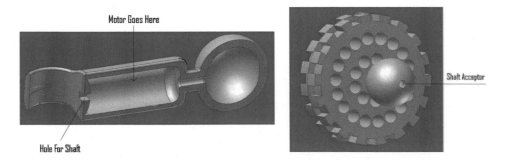

Fig. 11.1   A piece from a toy concept for improving an existing commercial toy—ZOOB [1]

Fig. 11.2 Original ZOOB components at work [1]

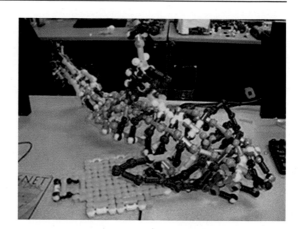

Fig. 11.3 Concept design of multiple toy pieces and their assemblies [1]

manufacturing of custom toys. A combination of rods and connectors for obtaining the final form of a possible design is given in Fig. 11.7.

11.2   Reverse Engineering Project

Student teams working on the product development project are also tasked to complete a reverse engineering project. Often the teams are supplied a toy to reverse engineer, while other times they are asked to find their own toys to analyze. Following is a list of the project requirements for such an effort. The teams have to follow the steps given below during the reverse engineering process, and record the results for their reporting in an MS PowerPoint presentation:

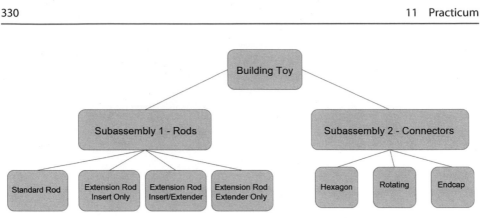

Fig. 11.4 System level design of a modular construction toy

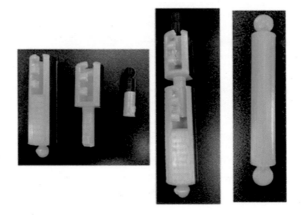

Fig. 11.5 Rod category parts: extension rod components and their assembly, and standard rod

- Find a toy or a product with no more than 10 parts (You could treat subassemblies of a system as a complete product or machine).
- Test the system in concern, to see if it is capable of its claims. Please elaborate about its performance, and its value (including its cost to the customers).
- Develop the black box model of the system with its inputs and outputs clearly identified (Fig. 11.8).
- Develop the glass box model of the system (similar to the system level design chart or an assembly chart—Fig. 11.9). Add an exploded assembly photograph of the system to complete this requirement (Fig. 11.9).
- Identify the materials and manufacturing processes involved in making the parts within the glass box model.
- Re-engineer the system by recommending design and performance (functionality) improvements.

Fig. 11.6 Connectors: hexagon, end cap, and rotating, along with an example of subassembly 2 with multiple rotating connectors

Fig. 11.7 Assembly of rods and connectors

This assignment depicted above is about analyzing an existing design; however, the author also emphasizes re-engineering the reverse engineered toy (with the last requirement)—adding a synthesis component to the experience along with analysis.

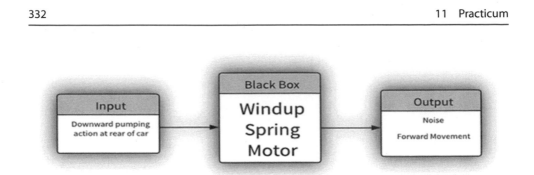

Fig. 11.8 Black box model of the toy car—Dinoco Cruz Ramirez [2]

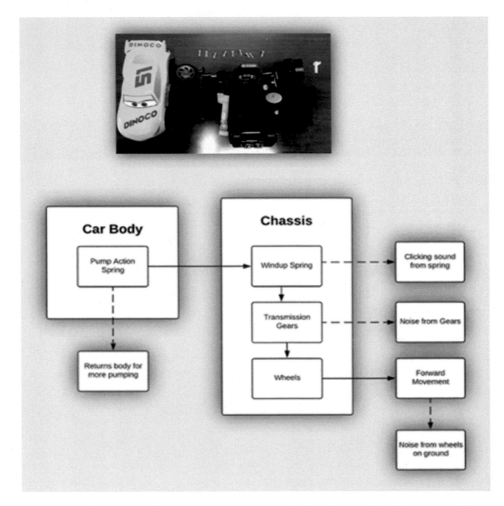

Fig. 11.9 Exploded assembly photograph and associated Glass Box model of the toy car—Dinoco Cruz Ramirez [2]

11.3   Automating Additive Manufacturing

In addition to the projects mentioned in the previous sections of this chapter, a new type of project may become prevalent. The author recently worked with two groups of his students. They not only developed light shades (diffusors) to be 3D printed for a company, but also conceptualized an automated 3D printing environment for meeting the demand of additively manufacturing large quantities of low-cost lighting elements. The Raise 2 Pro 3D system was originally selected due to its large build volume. However, Ultimaker S5 Pro Bundle was chosen due to the quality of the prints as well as the availability of a large number of spools for longevity of the printing process to meet the capacity requirements. A system like S5 Pro Bundle can be interfaced with a supervisor computer that may schedule or shift a print to another machine within the same print cluster or farm (*Printing Operating/Management Systems—i.e. 3DPrinterOS by 3D Systems or HP PrintOS*). 3D printer management software is becoming a focus with added workflow components as well [3].

References

1. Sirinterlikci, A., (2016). Open-Ended Design Projects in a Rapid Prototyping Course. American Society for Engineering Education (ASEE) Annual Conference and Exposition—Manufacturing Division, New Orleans, LA.
2. Sirinterlikci, A., Mativo, J., (2019). Critical Thinking in Manufacturing Engineering Education, 2019 ASEE Annual (American Society for Engineering Education) Conference and Exposition—Manufacturing Division, Tampa, FL.
3. Lavi, G., (2022). Track Your Prints—Top 3D Printer Farm & AM Workflow Management Software Solutions, located at https://all3dp.com/1/3d-printing-workflow-mes-software-buyers-guide/, accessed September 20, 2022.

Appendix I: G&M Codes

Reproduced from ANSI/EIA Standard, RS-274-D.

RS-274-D
Page 14

APPENDIX B

PREPARATORY AND MISCELLANEOUS FUNCTIONS

B.1   PREPARATORY FUNCTIONS (For Explanation, see Section B.3)

Code	Function Retained Until Cancelled or Superceded by Subsequent Command of the Same Letter Designation	Function Affects Only the Block Within Which It Appears	Function
G00	A		Point to Point, Positioning
G01	A		Linear Interpolation
G02	A		Circular Interpolation Arc CW (2 Dimensional)
G03	A		Circular Interpolation Arc CCW (2 Dimensional)
G04		X	Dwell
G05	*	*	Unassigned
G06	A		Parabolic Interpolation
G07	*	*	Unassigned
G08		X	Acceleration
G09		X	Deceleration
G10-G12	*	*	Unassigned
G13-G16	B		Axis Selection
G17	C		XY Plane Selection
G18	C		ZX Plane Selection
G19	C		YZ Plane Selection
G20-G24	*	*	Unassigned
G25-G29	*	*	Permanently Unassigned
G30-G32	*	*	Unassigned
G33	A		Threadcutting, Constant Lead
G34	A		Threadcutting, Increasing Lead
G35	A		Threadcutting, Decreasing Lead
G36-G39	*	*	Permanently Unassigned
G40	D		Cutter Compensation/Offset, Cancel
G41	D		Cutter Compensation-Left
G42	D		Cutter Compensation-Right
G43	D		Cutter Offset, Inside Corner
G44	D		Cutter Offset, Outside Corner
G45-G49	*	*	Unassigned
G50-G59	*	*	Reserved for Adaptive Control
G60-G69	*	*	Unassigned
G70	I		Inch Programming
G71	I		Metric Programming

RS-274-D
Page 15

B.1   (Cont.)

Code	Function Retained Until Cancelled or Superceded by Subsequent Command of the Same Letter Designation	Function Affects Only the Block Within Which It Appears	Function
G72	A		Circular Interpolation — CW (3 Dimensional)
G73	A		Circular Interpolation — CCW (3 Dimensional)
G74	J		Cancel Multiquadrant Circular Interpolation
G75	J		Multiquadrant Circular Interpolation
G76-G79	*	*	Unassigned
G80	E		Fixed Cycle Cancel
G81	E		Fixed Cycle No. 1
G82	E		Fixed Cycle No. 2
G83	E		Fixed Cycle No. 3
G84	E		Fixed Cycle No. 4
G85	E		Fixed Cycle No. 5
G86	E		Fixed Cycle No. 6
G87	E		Fixed Cycle No. 7
G88	E		Fixed Cycle No. 8
G89	E		Fixed Cycle No. 9
G90	F		Absolute Dimension Input
G91	F		Incremental Dimension Input
G92		X	Preload Registers
G93	G		Inverse Time Feedrate (V/D)
G94	G		Inches (millimeters) per Minute Feedrate
G95	G		Inches (millimeters) per Spindle Revolution
G96	H		Constant Surface Speed, Feet (meters) per Minute
G97	H		Revolutions per Minute
G98-G99	*	*	Unassigned

*The choice of a particular case must be designated in the Format Classification Sheet. (See Appendix A.)

NOTES:

1. Permanently unassigned codes are for individual use and are not intended to be assigned in the next revision of the standards.

2. For future selection of "G" code pairs for on-off functions, the higher number shall initiate the function and the lower number shall cancel the function.

3. Assignments in previous revision: G72, G73, G74, G75 unassigned.

RS-274-D
Page 16

B.2   MISCELLANEOUS FUNCTIONS (For Explanations, see Section B.3)

Code	Function Starts Relative to Commanded Motion in Its Block		Function Retained Until Cancelled or Superceded by an Appropriate Subsequent Command	Function Affects Only the Block Within It Appears	Function
	With	After Completion			
M00		X		X	Program Stop
M01		X		X	Optional (Planned) Stop
M02		X		X	End of Program
M03	X		X		Spindle CW
M04	X		X		Spindle CCW
M05		X	X		Spindle OFF
M06	*	*		X	Tool Change
M07	X		X		Coolant No. 2 ON
M08	X		X		Coolant No. 1 ON
M09		X	X		Coolant OFF
M10	*	*	X		Clamp
M11	*	*	X		Unclamp
M12		X		X	Synchronization Code
M13	X		X		Spindle CW & Coolant ON
M14	X		X		Spindle CCW & Coolant ON
M15	X			X	Motion +
M16	X			X	Motion −
M17-M18	*	*	*	*	Unassigned
M19	*	*	X		Oriented Spindle Stop
M20-M29	*	*	*	*	Permanently Unassigned
M30		X		X	End of Data
M31	X			X	Interlock Bypass
M32-M35	*	*	*	*	Unassigned
M36-M39	*	*	*	*	Permanently Unassigned
M40-M46	X		X		Gear Changes if Used; Otherwise Unassigned
M47	*	*	*	*	Return to Program Start
M48	X		X		Cancel M49
M49	X		X		Bypass Override
M50-M57	*	*	*	*	Unassigned
M58	X		X		Cancel M59
M59	X		X		Bypass CSS Updating
M60-M89	*	*	*	*	Unassigned
M90-M99	*	*	*	*	Reserved for User

*The choice of a particular case must be designated in the Format Classification Sheet. (See Appendix A.)

NOTES:

1. Permanently unassigned codes are for individual use and are not intended to be assigned in the next revision of the standard.
2. For future selection of M codes, the higher number shall initiate the function and the lower number cancel it.
3. Assignments in previous revision: M12, M46, M47, M58, M59 Unassigned.

RS-274-D
Page 17

B.3   EXPLANATIONS OF FUNCTIONS

G00	Point to Point Positioning	Point to point positioning at rapid or other traverse rate.*
G01	Linear Interpolation	A mode of contouring control which uses the information contained in a block to produce a straight line in which the vectorial velocity is held constant.
G02	Arc Clockwise (2 Dimensional)	An arc generated by the coordinated motion of two axes in which curvature of the path of the tool with respect to the workpiece is clockwise, when viewing the plane of motion in the negative direction of the perpendicular axis.
G03	Arc Counterclockwise (2 Dimensional)	An arc generated by the coordinated motion of two axes in which curvature of the path of the tool with respect to the workpiece is counterclockwise, when viewing the plane of motion in the negative direction of the perpendicular axis.
G02-G03	Circular Interpolation (2 Dimensional)	A mode of contouring control which uses the information contained in a single block to produce an arc of a circle. The velocities of the axes used to generate this arc are varied by the control.*
G04	Dwell	A timed delay of programmed or established duration, not cyclic or sequential; i.e., not an interlock or hold.
G06	Parabolic Interpolation	A mode of interpolation used in contouring to produce a segment of a parabola. Velocities of the axes used to generate this curve are varied by the control.
G08	Acceleration	A controlled velocity increase to programmed rate starting immediately.
G09	Deceleration	A controlled velocity decrease to a fixed percent of the programmed rate starting immediately.
G13-G16	Axis Selection	Used to direct a control to the axis or axes, as specified by the Format Classification, as in a system which time-shared the controls.*
G17-G19	Plane Selection	Used to identify the plane for such functions as Circular Interpolation, Cutter Compensation, and others as required.
G33	Thread Cutting, Constant Lead	Mode selection for machines equipped for thread cutting.
G34	Thread Cutting, Increasing Lead	Mode selection for machines equipped for thread cutting where a constantly increasing lead is desired.
G35	Threat Cutting, Decreasing Lead	Mode selection for machines equipped for thread cutting where a constantly decreasing lead is desired.

RS-274-D
Page 18

B.3   (Cont.)

G40	Cutter Compensation/Offset Cancel	Command which will discontinue any cutter comsation/offset
G41	Cutter Compensation-Left	Cutter on left side of work surface looking from cutter in the direction of relative cutter motion with displacement normal to the cutter path to adjust for the difference between actual and programmed cutter radii or diameters.
G42	Cutter Compensation-Right	Cutter on right side of work surface looking from cutter in the direction of relative cutter motion with displacement normal to the cutter path to adjust for the difference between actual and programmed cutter radii or diameters.
G43	Cutter Offset-Inside Corner	Displacement normal to cutter path to adjust for the difference between actual and programmed cutter radii or diameters. Cutter on inside corner.
G44	Cutter Offset-Outside Corner	Displacement normal to cutter path to adjust for the difference between actual and programmed cutter radii or diameters. Cutter on outside corner.
G50-G59	Adaptive Control	Reserved for adaptive control requirements.
G70	Inch Programming	Mode for programming in inch units. It is recommended that control turn on establish this mode of operation.
G71	Metric Programming	Mode for programming in metric units. This mode is cancelled by G70, M02, and M30.
G72	Arc Clockwise (3 Dimensional)	An arc generated by the coordinated motion of 3 axes in which the curvature of the tool path with respect to the workpiece is clockwise when viewed as described in Paragraph 7.2.
G73	Arc Counterclockwise (3 Dimensional)	An arc generated by the coordinated motion of 3 axes in which the curvature of the tool path with respect to the workpiece is counterclockwise when viewed as described in Paragraph 7.2.
G72-G73	Circular Interpolation (3 Dimensional)	A mode of contouring control which uses the information contained in a single block to produce an arc on a sphere. The velocities of the axes used to generate this arc are varied by the control.*
G75	Multi-Quadrant Circular	MODE Selection if required for Multi-Quadrant Circular, cancelled by G74.
G80		Command that will discontinue any of the fixed cycles G81-G89.

RS-274-D
Page 19

B.3   (Cont.)

G81-G89   Fixed Cycle**   A preset series of operations which direct machine axis movement and/or cause spindle operation to complete such action as boring, drilling, tapping, or combinations thereof.

| Fixed Cycle | | | At Bottom | | | |
Number	Code	Movement In	Dwell	Spindle	Movement Out to Feed Start	Typical Usage
1	G81	Feed	— —	— —	Rapid	Drill, Spot Drill
2	G82	Feed	Yes	— —	Rapid	Drill, Counterbore
3	G83	Intermittent	— —	— —	Rapid	Deep Hole
4	G84	Spindle Forward Feed	— —	Rev.	Feed	Tap
5	G85	Feed	— —	— —	Feed	Bore
6	G86	Start Spindle, Feed	— —	Stop	Rapid	Bore
7	G87	Start Spindle, Feed	—	Stop	Manual	Bore
8	G88	Start Spindle, Feed	Yes	Stop	Manual	Bore
9	G89	Feed	Yes	— —	Feed	Bore

G90   Absolute Input   A control mode in which the data input is in the form of absolute dimensions.

G91   Incremental Input   A control mode in which the data input is in the form of incremental data.

G92   Preload of Registers   Used to preload registers to desired values. No machine operation is initiated. Examples would include preload of axis position registers, spindle speed constraints, initial radius, etc. Information within this block shall conform to the character assignments of Table 1.

G93   Inverse Time Feedrate   The data following the feedrate address is equal to the reciprocal of the time in minutes to execute the blocks and is equivalent to the velocity of any axis divided by the corresponding programmed increment.

G94   Inches (Millimeters) Per Minute Feedrate   The feedrate code units are inches per minute or millimeters per minute.

G95   Inches (Millimeters) Per Revolution   The feedrate code units are inches (millimeters) per revolution of the spindle.

G96   Constant Surface Speed Per Minute   The spindle speed code units are surface feet (meters) per minute and specify the tangential surface speed of the tool relative to the workpiece. The spindle speed is automatically controlled to maintain the programmed value.

RS-274-D
Page 20

B.3   (Cont.)

G97	Revolutions Per Minute	The spindle speed is defined by the spindle speed word.
M00	Program Stop	A miscellaneous function command to cancel the spindle and coolant functions and terminate further program execution after completion of other commands in the block.
M01	Optional (Planned) Stop	A miscellaneous function command similar to a program stop except that the control ignores the command unless the operator has previously validated the command.*
M02	End of Program	A miscellaneous function indicating completion of workpiece. Stops spindle, coolant, and feed after completion of all commands in the block. Used to reset control and/or machine. Resetting control may include rewind of tape to the end of record character or progressing a loop tape through the splicing leader.*
M03	Spindle CW	Start spindle rotation to advance a right-handed screw into the workpiece.
M04	Spindle CCW	Start spindle rotation to retract a right-handed screw from the workpiece.
M05	Spindle Off	Stop spindle in normal, most efficient manner; brake, if available, applied; coolant turned off.
M07-M08-M09	Coolant, On, Off	Mist (No. 2), flood (No. 1), tapping coolant or dust collector.*
M10-M11	Clamp, Unclamp	Can pertain to machine slides, workpiece, fixtures, spindle, etc.
M12	Synchronization Code	An inhibiting code used for synchronization of multiple sets of axes. For example, see Appendix C.1.
M15-M16	Motion +, Motion −	Rapid Traverse or feed direction selection, where required.*
M19	Oriented Spindle Stop	A miscellaneous function which causes the spindle to stop at a predetermined or programmed angular position.*
M30	End of Data	A miscellaneous function which stops spindle, coolant, and feed after completion of all commands in the block. Used to reset control and/or machine. Resetting control will include rewind of tape to the end of record character, progressing a loop tape through the splicing leader, or transferring to a second tape reader.*

RS-274-D
Page 21

B.3   (Cont.)

M31	Interlock By-Pass	A command to temporarily circumvent a normally provided interlock.*
M47	Return to Program Start	A miscellaneous function which continues program execution from the start of program, unless inhibited by an interlock signal.
M49	Override By-Pass	A function which deactivates a manual spindle or feed override and returns the parameter to the programmed value. Cancelled by M48.
M59	CSS By-Pass Updating	A function which holds the RPM constant at its value when M59 is initiated/cancelled by M58.
M90-M99	Reserved for User	Miscellaneous function outputs which are reserved exclusively for the machine user.

*The choice for a particular case must be defined in the Format Classification Sheet. (See Appendix A.)

**This command initiates a sequence of events which will be repeated at the appropriate times until cancelled or changed.

NOTE: Additional commands may be required on specific machines. Unassigned code numbers should be used for these and specified on the Format Classification Sheet. On certain machines the functions described may not be completely applicable; deviations and interpretations should be clarified in the Format Classification Sheet. (See Appendix A.)

Appendix II: G&M Codes for 3D Printing

Taken from https://reprap.org/wiki/G-code.
- G-commands
- 11.1G0 & G1: Move
- 11.2G2 & G3: Controlled Arc Move
- 11.3G4: Dwell
- 11.4G6: Direct Stepper Move
- 11.5G10: Set tool Offset and/or workplace coordinates and/or tool temperatures
- 11.6G10: Retract
- 11.7G11: Unretract
- 11.8G12: Clean Tool
- 11.9G17…19: Plane Selection (CNC specific)
- 11.10G20: Set Units to Inches
- 11.11G21: Set Units to Millimeters
- 11.12G22: Firmware Retract
- 11.13G23: Firmware Recover
- 11.14G26: Mesh Validation Pattern
- 11.15G27: Park toolhead
- 11.16G28: Move to Origin (Home)
- 11.17G29: Detailed Z-Probe
- 11.17.1G29: Auto Bed Leveling (Marlin–MK4duo)
- 11.17.2G29: Unified Bed Leveling (Marlin–MK4duo)
- 11.17.3G29: Manual Bed Leveling (Marlin–MK4duo)
- 11.17.4G29: Auto Bed Leveling (Repetier-Firmware)
- 11.17.5G29: Mesh Bed Compensation (RepRapFirmware)
- 11.18G29.1: Set Z probe head offset
- 11.19G29.2: Set Z probe head offset calculated from toolhead position
- 11.20G30: Single Z-Probe
- 11.21G31: Set or Report Current Probe status
- 11.22G31: Dock Z Probe sled

© The Editor(s) (if applicable) and The Author(s), under exclusive license   345
to Springer Nature Switzerland AG 2023
A. Sirinterlikci and Y. Ertekin, *A Comprehensive Approach to Digital Manufacturing*,
Synthesis Lectures on Mechanical Engineering,
https://doi.org/10.1007/978-3-031-25354-6

- 11.23G32: Probe Z and calculate Z plane
- 11.23.1: Probe and calculate in Reprapfirmware
- 11.23.2: Probe and calculate in Repetier firmware
- 11.24G32: Undock Z Probe sled
- 11.25G33: Firmware dependent
- 11.25.1G33: Measure/List/Adjust Distortion Matrix (Repetier–Rcdeem)
- 11.25.2G33: Delta Auto Calibration (Marlin 1.1.x–MK4duo)
- 11.26G34: Z Stepper Auto-Align
- 11.27G34: Calculate Delta Height from toolhead position (DELTA)
- 11.28G38.x: Straight Probe (CNC specific)
- 11.28.1G38.2: Probe toward workpiece, stop on contact, signal error if failure
- 11.28.2G38.3: Probe toward workpiece, stop on contact
- 11.28.3G38.4: Probe away from workpiece, stop on loss of contact, signal error if failure
- 11.28.4G38.5: Probe away from workpiece, stop on loss of contact
- 11.29G40: Compensation Off (CNC specific)
- 11.30G42: Move to Grid Point
- 11.31G53...59: Coordinate System Select (CNC specific)
- 11.32G60: Save current position to slot
- 11.33G68: Coordinate rotation
- 11.34G69: Cancel coordinate rotation
- 11.35G75: Print temperature interpolation
- 11.36G76: PINDA probe temperature calibration
- 11.37G80: Cancel Canned Cycle (CNC specific)
- 11.38G80: Mesh-based Z probe
- 11.39G81: Mesh bed leveling status
- 11.40G82: Single Z probe at current location
- 11.41G83: Babystep in Z and store to EEPROM
- 11.42G84: UNDO babystep Z (move Z axis back)
- 11.43G85: Pick best babystep
- 11.44G86: Disable babystep correction after home
- 11.45G87: Enable babystep correction after home
- 11.46G88: Reserved
- 11.47G90: Set to Absolute Positioning
- 11.48G91: Set to Relative Positioning
- 11.49G92: Set Position
- 11.49.1G92.x: Reset Coordinate System Offsets (CNC specific)
- 11.50G93: Feed Rate Mode (Inverse Time Mode) (CNC specific)
- 11.51G94: Feed Rate Mode (Units per Minute) (CNC specific)
- 11.52G98: Activate farm mode
- 11.53G99: Deactivate farm mode

- 11.54G100: Calibrate floor or rod radius
- 11.55G130: Set digital potentiometer value
- 11.56G131: Remove offset
- 11.57G132: Calibrate endstop offsets
- 11.58G133: Measure steps to top
- 11.59G161: Home axes to minimum
- 11.60G162: Home axes to maximum
- 11.61G425: Perform auto-calibration with calibration cube
- 12M-commands
- 12.1M0: Stop or Unconditional stop
- 12.2M1: Sleep or Conditional stop
- 12.3M2: Program End
- 12.4M3: Spindle On, Clockwise (CNC specific)
- 12.5M4: Spindle On, Counter-Clockwise (CNC specific)
- 12.6M5: Spindle Off (CNC specific)
- 12.7M6: Tool change
- 12.8M7: Mist Coolant On (CNC specific)
- 12.9M8: Flood Coolant On (CNC specific)
- 12.10M9: Coolant Off (CNC specific)
- 12.11M10: Vacuum On (CNC specific)
- 12.12M11: Vacuum Off (CNC specific)
- 12.13M13: Spindle on (clockwise rotation) and coolant on (flood)
- 12.14M16: Expected Printer Check
- 12.15M17: Enable/Power all stepper motors
- 12.16M18: Disable all stepper motors
- 12.17M20: List SD card
- 12.18M21: Initialize SD card
- 12.19M22: Release SD card
- 12.20M23: Select SD file
- 12.21M24: Start/resume SD print
- 12.22M25: Pause SD print
- 12.23M26: Set SD position
- 12.24M27: Report SD print status
- 12.25M28: Begin write to SD card
- 12.26M29: Stop writing to SD card
- 12.27M30: Delete a file on the SD card
- 12.27.1M30: Program Stop
- 12.28M31: Output time since last M109 or SD card start to serial
- 12.29M32: Select file and start SD print
- 12.30M33: Get the long name for an SD card file or folder
- 12.31M33: Stop and Close File and save restart.gcode

- 12.32M34: Set SD file sorting options
- 12.33M35: Upload firmware NEXTION from SD
- 12.34M36: Return file information
- 12.35M36.1: Return embedded thumbnail data
- 12.36M37: Simulation mode
- 12.37M38: Compute SHA1 hash of target filc
- 12.38M39: Report SD card information
- 12.39M40: Eject
- 12.40M41: Loop
- 12.41M42: Switch I/O pin
- 12.42M43: Stand by on material exhausted
- 12.43M43: Pin report and debug
- 12.44M44: Codes debug—report codes available
- 12.45M44: Reset the bed skew and offset calibration
- 12.46M45: Bed skew and offset with manual Z up
- 12.47M46: Show the assigned IP address
- 12.48M47: Show end stops dialog on the display
- 12.49M48: Measure Z-Probe repeatability
- 12.50M49: Set G26 debug flag
- 12.51M70: Display message
- 12.52M72: Play a tone or song
- 12.53M73: Set/Get build percentage
- 12.54M75: Start the print job timer
- 12.55M76: Pause the print job timer
- 12.56M77: Stop the print job timer
- 12.57M78: Show statistical information about the print jobs
- 12.58M80: ATX Power On
- 12.59M81: ATX Power Off
- 12.60M82: Set extruder to absolute mode
- 12.61M83: Set extruder to relative mode
- 12.62M84: Stop idle hold
- 12.63M85: Set Inactivity Shutdown Timer
- 12.64M86: Set Safety Timer expiration time
- 12.65M87: Cancel Safety Timer
- 12.66M92: Set axis_steps_per_unit
- 12.67M93: Send axis_steps_per_unit
- 12.68M98: Call Macro/Subprogram
- 12.69M99: Return from Macro/Subprogram
- 12.70M101: Turn extruder 1 on (Forward), Undo Retraction
- 12.71M102: Turn extruder 1 on (Reverse)
- 12.72M102: Configure Distance Sensor

- 12.73M103: Turn all extruders off, Extruder Retraction
- 12.74M104: Set Extruder Temperature
- 12.74.1M104 in Marlin Firmware
- 12.74.2M104 in Teacup Firmware
- 12.74.3M104 in RepRapFirmware and Klipper
- 12.75M105: Get Extruder Temperature
- 12.76M106: Fan On
- 12.76.1M106 in RepRapFirmware
- 12.76.2M106 in Teacup Firmware
- 12.77M107: Fan Off
- 12.78M108: Cancel Heating
- 12.79M108: Set Extruder Speed (BFB)
- 12.80M109: Set Extruder Temperature and Wait
- 12.80.1M109 in Teacup
- 12.80.2M109 in Marlin, MK4duo, Sprinter (ATmega port), RepRapFirmware, Prusa
- 12.80.3M109 in Sprinter (4pi port)
- 12.80.4M109 in MakerBot
- 12.80.5M109 in Klipper
- 12.81M110: Set Current Line Number
- 12.82M111: Set Debug Level
- 12.82.1M111 in RepRapFirmware
- 12.82.2M111 in Repetier
- 12.83M112: Full (Emergency) Stop
- 12.84M113: Set Extruder PWM
- 12.85M113: Host Keepalive
- 12.86M114: Get Current Position
- 12.87M115: Get Firmware Version and Capabilities
- 12.88M116: Wait
- 12.89M117: Get Zero Position
- 12.90M117: Display Message
- 12.91M118: Echo message on host
- 12.92M118: Negotiate Features
- 12.93M119: Get Endstop Status
- 12.94M120: Push
- 12.95M121: Pop
- 12.96M120: Enable endstop detection
- 12.97M121: Disable endstop detection
- 12.98M122: Firmware dependent
- 12.98.1M122: Diagnose (RepRapFirmware)
- 12.98.2M122: Set Software Endstop (MK4duo)
- 12.98.3M122: Debug Stepper drivers (Marlin)

- 12.99M123: Firmware dependent
- 12.99.1M123: Tachometer value (RepRap, Prusa and Marlin)
- 12.99.2M123: Endstop Logic (MK4duo)
- 12.100M124: Firmware dependent
- 12.100.1M124: Immediate motor stop
- 12.100.2M124: Set Endstop Pullup
- 12.101M126: Open Valve
- 12.101.1M126 in MakerBot
- 12.102M127: Close Valve
- 12.102.1M127 in MakerBot
- 12.103M128: Extruder Pressure PWM
- 12.104M129: Extruder pressure off
- 12.105M130: Set PID P value
- 12.106M131: Set PID I value
- 12.107M132: Set PID D value
- 12.107.1M132 in MakerBot
- 12.108M133: Set PID I limit value
- 12.108.1M133 in MakerBot
- 12.109M134: Write PID values to EEPROM
- 12.109.1M134 in MakerBot
- 12.110M135: Set PID sample interval
- 12.110.1M135 in MakerBot
- 12.111M136: Print PID settings to host
- 12.112M140: Set Bed Temperature (Fast)
- 12.113M141: Set Chamber Temperature (Fast)
- 12.114M142: Firmware dependent
- 12.114.1M142: Holding Pressure
- 12.114.2M142: Set Cooler Temperature (Fast)
- 12.115M143: Maximum heater temperature
- 12.116M144: Bed Standby
- 12.117M146: Set Chamber Humidity
- 12.118M149: Set temperature units
- 12.119M150: Set LED color
- 12.120M154: Auto Report Position
- 12.121M155: Automatically send temperatures
- 12.122M160: Number of mixed materials
- 12.123M163: Set weight of mixed material
- 12.124M164: Store weights
- 12.125M165: Set multiple mix weights
- 12.126M190: Wait for bed temperature to reach target temp
- 12.127M191: Wait for chamber temperature to reach target temp

- 12.128M192: Wait for Probe Temperature
- 12.129M200: Set filament diameter
- 12.130M201: Set max acceleration
- 12.131M201.1: Set reduced acceleration for special move types
- 12.132M202: Set max travel acceleration
- 12.133M203: Firmware dependent
- 12.133.1M203: Set maximum feedrate
- 12.133.2M203 (Repetier): Set temperature monitor
- 12.134M204: Firmware dependent
- 12.134.1M204: Set default acceleration
- 12.134.2M204 (Repetier): Set PID values
- 12.135M205: Firmware dependent
- 12.135.1M205: Advanced settings
- 12.135.2M205 (Repetier): EEPROM Report
- 12.136M206: Firmware dependent
- 12.136.1M206: Offset axes
- 12.136.2M206 (Repetier): Set EEPROM value
- 12.137M207: Firmware dependent
- 12.137.1M207: Set retract length
- 12.137.2M207: Calibrate Z axis with Z max endstop
- 12.137.3M207 (Repetier): Set jerk without saving to EEPROM
- 12.138M208: Firmware dependent
- 12.138.1M208: Set unretract length
- 12.138.2M208 (RepRapFirmware): Set axis max travel
- 12.139M209: Enable automatic retract
- 12.140M210: Set homing feed rates
- 12.141M211: Disable/Enable software endstops
- 12.142M212: Set Bed Level Sensor Offset
- 12.143M217: Tool change Parameters
- 12.144M218: Set Hotend Offset
- 12.145M220: Set speed factor override percentage
- 12.146M221: Set extrude factor override percentage
- 12.147M220: Turn off AUX V1.0.5
- 12.148M221: Turn on AUX V1.0.5
- 12.149M222: Set speed of fast XY moves
- 12.150M223: Set speed of fast Z moves
- 12.151M224: Enable extruder during fast moves
- 12.152M225: Disable on extruder during fast moves
- 12.153M226: G-code Initiated Pause
- 12.154M226: Wait for pin state
- 12.155M227: Enable Automatic Reverse and Prime

- 12.156M228: Disable Automatic Reverse and Prime
- 12.157M229: Enable Automatic Reverse and Prime
- 12.158M230: Disable/Enable Wait for Temperature Change
- 12.159M231: Set OPS parameter
- 12.160M232: Read and reset max. advance values
- 12.161M240: Trigger camera
- 12.162M240: Start conveyor belt motor/Echo off
- 12.163M241: Stop conveyor belt motor/echo on
- 12.164M245: Start cooler
- 12.165M246: Stop cooler
- 12.166M250: Set LCD contrast
- 12.167M256: Set LCD brightness
- 12.168M251: Measure Z steps from homing stop (Delta printers)
- 12.169M260: i2c Send Data
- 12.170M261: i2c Request Data
- 12.171M280: Set servo position
- 12.172M281: Set Servo Angles
- 12.173M282: Detach Servo
- 12.174M290: Babystepping
- 12.175M291: Display message and optionally wait for response
- 12.176M292: Acknowledge message
- 12.177M300: Play beep sound
- 12.178M301: Set PID parameters
- 12.178.1: MK4duo
- 12.178.2: Marlin
- 12.178.3: RepRapFirmware 1.15 onwards
- 12.178.4: RepRapFirmware 1.09 to 1.14 inclusive
- 12.178.5: Smoothie
- 12.178.6: Other implementations
- 12.178.7: Teacup
- 12.179M302: Allow cold extrudes
- 12.180M303: Run PID tuning
- 12.181M304: Set PID parameters—Bed
- 12.181.1M304 in RepRapPro version of Marlin: Set thermistor values
- 12.182M305: Set thermistor and ADC parameters
- 12.183M306: Set home offset calculated from toolhead position
- 12.184M307: Set or report heating process parameters
- 12.185M308: Set or report sensor parameters
- 12.186M309: Set or report heater feedforward
- 12.187M310: Temperature model settings
- 12.188M320: Activate autolevel (Repetier)

- 12.189M321: Deactivate autolevel (Repetier)
- 12.190M322: Reset autolevel matrix (Repetier)
- 12.191M323: Distortion correction on/off (Repetier)
- 12.192M340: Control the servos
- 12.193M350: Set microstepping mode
- 12.194M351: Toggle MS1 MS2 pins directly
- 12.195M355: Turn case lights on/off
- 12.196M360: Report firmware configuration
- 12.197SCARA calibration codes (Morgan)
- 12.198M360: Move to Theta 0 degree position
- 12.199M361: Move to Theta 90 degree position
- 12.200M362: Move to Psi 0 degree position
- 12.201M363: Move to Psi 90 degree position
- 12.202M364: Move to Psi + Theta 90 degree position
- 12.203M365: SCARA scaling factor
- 12.204M366: SCARA convert trim
- 12.205M370: Morgan manual bed level—clear map
- 12.206M371: Move to next calibration position
- 12.207M372: Record calibration value, and move to next position
- 12.208M373: End bed level calibration mode
- 12.209M374: Save calibration grid
- 12.210M375: Display matrix/Load Matrix
- 12.211M376: Set bed compensation taper
- 12.212M380: Activate solenoid
- 12.213M381: Disable all solenoids
- 12.214M400: Wait for current moves to finish
- 12.215M401: Deploy Z Probe
- 12.216M402: Stow Z Probe
- 12.217M403: Set filament type (material) for particular extruder and notify the MMU
- 12.218M404: Filament width and nozzle diameter
- 12.219M405: Filament Sensor on
- 12.220M406: Filament Sensor off
- 12.221M407: Display filament diameter
- 12.222M408: Report JSON-style response
- 12.223M409: Query object model
- 12.224M410: Quick-Stop
- 12.225M412: Disable Filament Runout Detection
- 12.226M413: Power-Loss Recovery
- 12.227M415: Host Rescue
- 12.228M416: Power loss
- 12.229M420: Set RGB Colors as PWM (MachineKit)

- 12.230M420: Leveling On/Off/Fade (Marlin)
- 12.231M421: Set a Mesh Bed Leveling Z coordinate
- 12.232M422: Set a G34 Point
- 12.233M423: X-Axis Twist Compensation
- 12.234M425: Backlash Correction
- 12.235M450: Report Printer Mode
- 12.236M451: Select FFF Printer Mode
- 12.237M452: Select Laser Printer Mode
- 12.238M453: Select CNC Printer Mode
- 12.239M460: Define temperature range for thermistor-controlled fan
- 12.240M470: Create Directory on SD-Card
- 12.241M471: Rename File/Directory on SD-Card
- 12.242M486: Cancel Object
- 12.243M500: Store parameters in non-volatile storage
- 12.244M501: Read parameters from EEPROM
- 12.245M502: Restore Default Settings
- 12.246M503: Report Current Settings
- 12.247M504: Validate EEPROM
- 12.248M505: Firmware dependent
- 12.248.1M505: Clear EEPROM and RESET Printer
- 12.248.2M505: Set configuration file folder
- 12.249M509: Force language selection
- 12.250M510: Lock Machine
- 12.251M511: Unlock Machine with Passcode
- 12.252M512: Set Passcode
- 12.253M524: Abort SD Printing
- 12.254M530: Enable printing mode
- 12.255M531: Set print name
- 12.256M532: Set print progress
- 12.257M540: Set MAC address
- 12.258M540 in Marlin: Enable/Disable "Stop SD Print on Endstop Hit"
- 12.259M550: Set Name
- 12.260M551: Set Password
- 12.261M552: Set IP address, enable/disable network interface
- 12.262M553: Set Netmask
- 12.263M554: Set Gateway and/or DNS server
- 12.264M555: Set compatibility
- 12.265M556: Axis compensation
- 12.266M557: Set Z probe point or define probing grid
- 12.267M558: Set Z probe type
- 12.268M559: Upload configuration file

- 12.269M560: Upload web page file
- 12.270M561: Set Identity Transform
- 12.271M562: Reset temperature fault
- 12.272M563: Define or remove a tool
- 12.273M564: Limit axes
- 12.274M565: Set Z probe offset
- 12.275M566: Set allowable instantaneous speed change
- 12.276M567: Set tool mix ratios
- 12.277M568: Tool settings
- 12.278M568: Turn off/on tool mix ratios (obsolete meaning in old RepRapFirmware versions)
- 12.279M569: Stepper driver control
- 12.280M569.1: Stepper driver closed loop configuration
- 12.281M569.2: Read or write any stepper driver register
- 12.282M569.3: Read Motor Driver Encoder
- 12.283M569.4: Set Motor Driver Torque Mode
- 12.284M569.5: Collect Data from Closed-loop Driver
- 12.285M569.6: Execute Closed-loop Driver Tuning Move
- 12.286M569.7: Configure motor brake port
- 12.287M569.8: Read Axis Force
- 12.288M569.9: Sets the driver sense resistor and maximum current
- 12.289M570: Configure heater fault detection
- 12.290M571: Set output on extrude
- 12.291M572: Set or report extruder pressure advance
- 12.292M573: Report heater PWM
- 12.293M574: Set endstop configuration
- 12.294M575: Set serial comms parameters
- 12.295M576: Set SPI comms parameters
- 12.296M577: Wait until endstop is triggered
- 12.297M578: Fire inkjet bits
- 12.298M579: Scale Cartesian axes
- 12.299M580: Select Roland
- 12.300M581: Configure external trigger
- 12.301M582: Check external trigger
- 12.302M584: Set drive mapping
- 12.303M585: Probe Tool
- 12.304M586: Configure network protocols
- 12.305M587: Store WiFi host network in list, or list stored networks
- 12.306M588: Forget WiFi host network
- 12.307M589: Configure access point parameters
- 12.308M590: Report current tool type and index

- 12.309M591: Configure filament monitoring
- 12.310M592: Configure nonlinear extrusion
- 12.311M593: Configure Dynamic Acceleration Adjustment
- 12.312M594: Enter/Leave Height Following mode
- 12.313M595: Set movement queue length
- 12.314M596: Select movement queue number
- 12.315M597: Collision avoidance
- 12.316M600: Set line cross section
- 12.317M600: Filament change pause
- 12.318M601: Pause print
- 12.319M602: Resume print
- 12.320M603: Stop print (Prusa i3)
- 12.321M603: Configure Filament Change
- 12.322M605: Set dual x-carriage movement mode
- 12.323M650: Set peel move parameters
- 12.324M651: Execute peel move
- 12.325M665: Set delta configuration
- 12.326M666: Set delta endstop adjustment
- 12.327M667: Select CoreXY mode
- 12.328M668: Set Z-offset compensations polynomial
- 12.329M669: Set kinematics type and kinematics parameters
- 12.330M670: Set IO port bit mapping
- 12.331M671: Define positions of Z lead screws or bed leveling screws
- 12.332M672: Program Z probe
- 12.333M673: Align plane on rotary axis
- 12.334M674: Set Z to center point
- 12.335M675: Find center of cavity
- 12.336M700: Level plate
- 12.337M701: Load filament
- 12.338M702: Unload filament
- 12.339M703: Configure filament
- 12.340M704: Filament/MMU related gcode in development (reserve)
- 12.341M705: Filament/MMU related gcode in development (reserve)
- 12.342M706: Filament/MMU related gcode in development (reserve)
- 12.343M707: Filament/MMU related gcode in development (reserve)
- 12.344M708: Filament/MMU related gcode in development (reserve)
- 12.345M709: Filament/MMU related gcode in development (reserve)
- 12.346M710: Firmware dependent
- 12.346.1M710: Controller Fan settings
- 12.346.2M710: Erase the EEPROM and reset the board
- 12.347M750: Enable 3D scanner extension

- 12.348M751: Register 3D scanner extension over USB
- 12.349M752: Start 3D scan
- 12.350M753: Cancel current 3D scanner action
- 12.351M754: Calibrate 3D scanner
- 12.352M755: Set alignment mode for 3D scanner
- 12.353M756: Shutdown 3D scanner
- 12.354M800: Fire start print procedure
- 12.355M801: Fire end print procedure
- 12.356M808: Set or Goto Repeat Marker
- 12.356.1M808 in Marlin 2.0.8
- 12.357M810-M819: G-code macros stored in memory or flash not filename
- 12.358M851: Set Z-Probe Offset
- 12.358.1M851 in Marlin 1.0.2
- 12.358.2M851 in Marlin 1.1.0
- 12.358.3M851 in Marlin 2.0.0
- 12.358.4M851 in MK4duo 4.3.25
- 12.358.5M851 in RepRapFirmware 2.02 and later
- 12.359M860 Wait for Probe Temperature
- 12.360M861 Set Probe Thermal Compensation
- 12.361M862: Print checking
- 12.361.1M862.1: Check nozzle diameter
- 12.361.2M862.2: Check model code
- 12.361.3M862.3: Model name
- 12.361.4M862.4: Firmware version
- 12.361.5M862.5: Gcode level
- 12.362M871: PTC Configuration
- 12.363M876: Dialog handling
- 12.364M890: Run User Gcode
- 12.365M900: Set Linear Advance Scaling Factors
- 12.366M905: Set local date and time
- 12.367M906: Set motor currents
- 12.368M907: Set digital trimpot motor
- 12.369M908: Control digital trimpot directly
- 12.370M909: Set microstepping
- 12.371M910: Set decay mode
- 12.372M910: TMC2130 init
- 12.373M911: Configure auto save on loss of power ("power panic")
- 12.374M911: Set TMC2130 holding currents
- 12.375M912: Set electronics temperature monitor adjustment
- 12.376M912: Set TMC2130 running currents
- 12.377M913: Set motor percentage of normal current

- 12.378M913: Print TMC2130 currents
- 12.379M914: Set/Get Expansion Voltage Level Translator
- 12.380M914: Set TMC2130 normal mode
- 12.381M915: Configure motor stall detection
- 12.382M915: Set TMC2130 silent mode
- 12.383M916: Resume print after power failure
- 12.384M916: Set TMC2130 Stallguard sensitivity threshold
- 12.385M917: Set motor standstill current reduction
- 12.386M917: Set TMC2130 PWM amplitude offset (pwm_ampl)
- 12.387M918: Configure direct-connect display
- 12.388M918: Set TMC2130 PWM amplitude gradient (pwm_grad)
- 12.389M928: Start SD logging
- 12.390M929: Start/stop event logging to SD card
- 12.391M950: Create heater, fan or GPIO/servo device
- 12.392M951: Set height following mode parameters
- 12.393M952: Set CAN expansion board address and/or normal data rate
- 12.394M953: Set CAN-FD bus fast data rate
- 12.395M954: Configure as CAN expansion board
- 12.396M955: Configure Accelerometer
- 12.397M956: Collect accelerometer data and write to file
- 12.398M957: Raise event
- 12.399M995: Calibrate Touch Screen
- 12.400M997: Perform in-application firmware update
- 12.401M998: Request resend of line
- 12.402M999: Restart after being stopped by error
- Other commands
- 13.1G: List all G-codes
- 13.2M: List all M-codes
- 13.3T: Select Tool
- 13.4D: Debug codes
- 13.4.1D-1: Endless Loop
- 13.4.2D0: Reset
- 13.4.3D1: Clear EEPROM and RESET
- 13.4.4D2: Read/Write RAM
- 13.4.5D3: Read/Write EEPROM
- 13.4.6D4: Read/Write PIN
- 13.4.7D5: Read/Write FLASH
- 13.4.8D6: Read/Write external FLASH
- 13.4.9D7: Read/Write Bootloader
- 13.4.10D8: Read/Write PINDA
- 13.4.11D9: Read/Write ADC

- 13.4.12D10: Set XYZ calibration = OK
- 13.4.13D12: Time
- 13.4.14D20: Generate an offline crash dump
- 13.4.15D21: Print crash dump to serial
- 13.4.16D22: Clear crash dump state
- 13.4.17D23: Request emergency dump on serial
- 13.4.18D80: Bed check
- 13.4.19D81: Bed analysis
- 13.4.20D106: Print measured fan speed for different pwm values
- 13.4.21D2130: Trinamic stepper controller

Appendix III: Purdue University AM Scorecard

Reproduced from Booth, J. W., Alperovich, J., Reid, T. N., Ramani, K. (2016), The Design for Additive Manufacturing Worksheet, Proceedings of the ASME International Design Engineering Technical Conferences and Computers and Information in Engineering Conference, IDETC/CIE 2016, Charlotte, NC.

© The Editor(s) (if applicable) and The Author(s), under exclusive license
to Springer Nature Switzerland AG 2023
A. Sirinterlikci and Y. Ertekin, *A Comprehensive Approach to Digital Manufacturing*,
Synthesis Lectures on Mechanical Engineering,
https://doi.org/10.1007/978-3-031-25354-6

Mark one	Complexity	Mark one	Functionality	Mark one	Material removal	Mark one	Unsupported features	Sum across rows		Totals
	Note: simple parts are inefficient for AM		Note: AM parts are light and medium duty		Note: support structures ruin surface finish		Note: unsupported features will droop			
☐ †	The part is the same shape as common stock materials, or is completely 2D	☐ *	Mating surfaces are bearing surfaces, or are expected to endure for 1000+ of cycles	☐	The part is smaller than or the same size as the required support structure	☐	There are long, unsupported features	0	x5	0
☐ *	The part is mostly 2D and can be made in a mill or lathe without repositioning it in the clamp	☐ *	Mating surfaces move significantly, experience large forces, or must endure 100–1000 cycles	☐	There are small gaps that will require support structures	☐	There are short, unsupported features	0	x4	0
☐	The part can be made in a mill or lathe, but only after repositioning it in the clamp at least once	☐	Mating surfaces move somewhat, experience moderate forces, or are expected to last 10–100 cycles	☐	Internal cavities, channels, or holes do not have openings for removing materials	☐	Overhang features have a slopped support	0	x3	0

Mark one	Complexity	Mark one	Functionality	Mark one	Material removal	Mark one	Unsupported features	Sum across rows		Totals
	Note: simple parts are ineffcient for AM		Note: AM parts are light and medium duty		Note: support structures ruin surface finish		Note: unsupported features will droop			
☐	The part curvature is complex(splines or arcs) for a machining operation such as a mill or lathe	☐	Mating surfaces will move minimally, experience low forces, or are intended to endure 2–10 cycles	☐	Material can be easily removed from internal cavities, channels, or holes	☐	Overhanging features have a minimum of 45° support	0	x2	0
☐	There are interior features or surface curvature is too complex to be machined	☐	Surfaces are purely non-functional or experience virtually no cycles	☐	There are no internal cavities, channels, or holes	☐	Part is oriented so there are no overhanging features	0	x1	0

Instructions: Mark one for each category for the part you plan to print. Check daggers and stars first, then scores

Mark one	Thin features	Mark one	Stress concentration	Mark one	Tolerances	Mark one	Geometric exactness			+
	Note: thin features will almost always break		Note: interior corners must transition gradually		Note: mating parts should not be the same size		Note: large flat areas tend to warp			
☐	Some Walls are less than 1/16" (1.5 mm) thick	☐	Interior corners have no chamfer, fillet, or rib	☐	Hole or length dimensions are nominal	☐	The part has large, flat surfaces or has a form that is important to be exact	0	x5	0
☐	Walls are between 1/16" (1.5 mm) and 1/8" (3 mm) thick	☐	Interior corners have chamfers, fillets, and/or ribs	☐	Hole or length tolerances are adjusted for shrinkage or fit	☐	The part has medium-sized, flat surfaces, or forms that are or should be close to exact	0	x3	0
☐	Walls are more than 1/8" (3 mm) thick	☐	Interior corners have generous chamfers, fillets, and/or ribs	☐	Hole and length tolerances are considered or are not important	☐	The part has small or no flat surfaces, or forms that need to be exact	0	x1	0
									Overall total	0

Starred ratings	
*	Consider a different manufacturing process
†	Stongly consider a different manufacturing process

Total score	
33–40	Needs redsign
24–32	Consider redesign
16–23	Moderate likelihood of success
8–15	Higher likelihood of success

Printed in the United States
by Baker & Taylor Publisher Services